The evolutionary ecology of ant–plant mutualisms

CAMBRIDGE STUDIES IN ECOLOGY

ALSO IN THE SERIES

Hugh G. Gauch, Jr. *Multivariate analysis in community ecology*
Robert Henry Peters *The ecological implications of body size*
C. S. Reynolds *The ecology of freshwater phytoplankton*
K. A. Kershaw *Physiological ecology of lichens*
Robert P. McIntosh *The background of ecology: concept and theory*

The evolutionary ecology of ant–plant mutualisms

ANDREW J. BEATTIE

Northwestern University

CAMBRIDGE UNIVERSITY PRESS

Cambridge
London New York New Rochelle
Sydney Melbourne

Published by the Press Syndicate of the University of Cambridge
The Pitt Building, Trumpington Street, Cambridge CB2 1RP
32 East 57th Street, New York, NY 10022, USA
10 Stamford Road, Oakleigh, Melbourne 3166, Australia

First published 1985

Printed in the United States of America

Library of Congress Cataloging in Publication Data

Beattie, Andrew J. (Andrew James), 1943–

The evolutionary ecology of ant–plant mutualisms.

(Cambridge studies in ecology)

Bibliography: p.

Includes index.

1. Ants – Ecology. 2. Ants – Evolution. 3. Insect–plant relationships. 4. Botany – Ecology. 5. Plants – Evolution. 6. Insects – Ecology. 7. Insects – Evolution. I. Title. II. Series.
QL568.F7B36 1985 595.79′6′0452482 84-27411
ISBN 0 521 25281 4 hard covers

To my mother and father,
Christine, and Helena

Contents

Preface

The natural history of ant–plant mutualisms has fascinated Western scientists for roughly two centuries. During this time it has become clear that the ways in which plants manipulate ants, and vice versa, can be so complex and subtle as to severely stretch the credence of the observer. The early natural historians described ant–plant relationships in superb detail, but generally inferred that a given relationship was mutualistic from anatomical, morphological, or behavioral data alone. Experimental verification was the exception rather than the rule. Although the experimental approach was tried by a few early workers, such as von Wettstein (1889), its impact was not dramatic until the publication of Janzen's seminal work on acacia ants about twenty years ago. This pioneering research has since been followed by many excellent experimental field studies embracing a variety of ant–plant mutualisms from many kinds of environments. Our knowledge of the selective pressures that produce the mutualistic response, the dynamics of the ant–plant interactions, the benefits to the plants and the ants, and the ways that mutualisms evolve has been vastly improved. At the same time, ant–plant mutualisms are extremely numerous and varied, and so far only a very few cases have been adequately analyzed. As a consequence, generalizations from limited data often have to be made. Whether or not this is wise will be revealed as new studies are published.

Ideas and syntheses generally enter the mind as a result of the stimulation of colleagues. I have been fortunate in having been at the receiving end of a great deal of stimulation, especially from Bob Abugov, Lin Chao, David Culver, Fran Hanzawa, Carol Horvitz, Dennis O'Dowd, Doug Schemske, Bob Taylor, and Christine Turnbull. Carol Horvitz, Doug Schemske, Dennis O'Dowd, and Herbert and Irene Baker have kindly allowed me to use some of their unpublished data. Comments and criticisms by Lin Chao, David Culver, Carol Horvitz, and Dennis O'Dowd of early drafts of various chapters have led to enormous improvements in the manuscript. I thank them for all the time and effort they put into their tasks, but urge the reader not to blame them for any errors that remain.

Tracy Ramsey typed the manuscript with great skill and patience. My wife, Christine Turnbull, made an excellent job of the illustrations and was also a constant and crucial source of encouragement.

1

Introduction

Walk through the vegetation almost anywhere on earth and you will see ants foraging on plants. The basic interaction has evolved into four ant–plant mutualisms in which ants: (1) protect plants from herbivores and other enemies, (2) feed plants essential nutrients, (3) disperse seeds and fruits, and (4) pollinate. Rewards produced by the plants, chiefly nest sites or food, are utilized by ants and the behavior patterns involved result in one or more of these services.

Much more is known about the benefits to the plants than the benefits to the ants. For example, the anatomy and morphology or chemistry of many of the rewards borne by the plants have been described. Critical field experiments have been performed to test the impact of ant services on plant growth, survivorship, and fecundity, and it is clear that ant services can profoundly affect plant fitness. On the other hand, although ants eagerly occupy plant-borne nest sites and harvest plant-borne food rewards, almost nothing precise is known about how this affects ant fitness. A major gap in our knowledge of ant–plant mutualisms is how the food rewards affect the physiology, growth, and demography of ant colonies. As a consequence of this situation, this book is written primarily from a "plant's point of view." It also should be noted that fundamentally nonmutualistic ant–plant interactions, such as predation by seed-gathering ants and herbivory by leaf-cutter ants, are referred to only in passing.

The origins and early evolution of ant–plant mutualisms are largely matters for speculation, but this has not stopped me from writing most of Chapter 2 on the subject. With the Cretaceous as a starting point, the characteristics of ant organization are discussed in the context of the kinds of selection pressure experienced by early angiosperms and other Cretaceous vegetation.

The mere presence of ant foragers on plants sometimes deters plant enemies such as herbivores. This theme is introduced in Chapter 2. In Chapter 3 it is shown that many plant species bear various kinds of ant attractants, most notably special cavities for nesting, small epidermal food bodies, and extrafloral nectar. Chapter 3 focuses on a discussion of the hypothesis that these attractants increase the frequency with which prey harmful to the plants is taken by ants. It is argued that selection for protection against enemies such as herbivores and seed predators has refined the attractants or rewards to the point where ant activity is sometimes sufficiently effective to reduce damage to plants and increase plant fitness. Chapter 4 concerns another ant attractant, "honeydew," which is secreted by a huge variety of homopterans and by some lepidopteran larvae. Although it is true that homopterans in particular are sophisiticated plant parasites, the possibility has been raised that when ants collect honeydew, their presence again constitutes a form of plant protection. This is a classic example of a three-way interaction which, if the ants and plants benefit, is an indirect mutualism mediated by the honeydew producers. These interactions become even more convoluted when both honeydew and extrafloral nectar are available on the same plant. Predicting the results of two-way, three-way, and more complex interactions has become an important aspect of evolutionary ecology. Ant–plant mutualisms provide some excellent examples of both simple and complex species interactions for empirical studies.

The highly specialized nesting cavities of a few plant species from the Far East and northern Australia are the focus of Chapter 5. Although the ants that occupy them may constitute a weak defense system, it is the nesting behavior itself that gives rise to the principal benefit to the plant. Waste materials abandoned by the ants within the nest cavities contain essential nutrients that the plants can absorb. In a real sense the ants feed the plants. A wide variety of other plant species bearing structures that encourage nesting by ants are included in the final discussion. Nests are commonly associated with these plants, but the services provided by the ants, if any, are frequently unknown.

Thousands of plant species produce fruits or seeds with special ant attractants called elaiosomes or arils. Ants eagerly harvest them, carrying the entire fruit or seed to the nest. There the attractants are removed and the seeds discarded with other wastes either in the nest, or outside a nest entrance. The ants feed the elaiosomes or arils principally to larvae. In Chapter 6 it is shown that seeds treated in this way are at a selective advantage because they are removed from the influence of some predators and, in fire-climax vegetation, from the effects of excessive heat. Seedlings that emerge from ant-manipulated seeds may suffer less competition at these microsites than others. However, the primary advantage to these seedlings is that they pass through the vulnerable establishment phase in a microsite enriched with ant waste materials that contain crucial, and often limiting, nutrients such as nitrogen and phosphorus. In a sense, ants relocate seeds from the point of release to private compost heaps.

Chapter 7 has a curious theme as ant pollination appears to be rare and the discussion revolves around the question as to why this should be. Only a few plant species are pollinated by ants, and in no case has a dependency on ants for seed set been proven beyond doubt. The hypothesis is introduced that ants secrete antibiotics inimical to pollen. The antibiotics have not evolved in response to the presence of pollen, but rather to the attacks of microorganisms. Ants may be particularly vulnerable to infection by bacteria and fungi as they commonly nest in cavities in the soil or decaying material such as fallen logs and dead twigs. In addition, unlike their relatives, the bees and the wasps, the juvenile stages are not enclosed in protective wax or paper cells. Pollen may resemble some microorganisms such as fungal spores in enough ways to suffer a similar fate when attacked by antibiotics.

The chemistry of food rewards for ants is outlined in Chapter 8. The quality, quantity, and timing of rewards are discussed in the context of colony physiology and demography. Although they often contain a variety of nutrients, the special importance of lipids is emphasized as this category of substances includes some insect dietary requirements. Any essential nutrient in plant-borne rewards that ants cannot themselves synthesize, and must locate by foraging, is potentially a powerful means of manipulating ant behavior.

In Chapter 9 variation in the ways ant–plant mutualisms function is attributed to a conglomerate of genetic, demographic, and ecological factors. These factors affect the outcome of interactions so that at any given location and time an ant–plant mutualism (a) may not be functioning at all, (b) may function weakly or erratically, or (c) may be operating to the advantage of some or all of the species involved. In this chapter the conditions that generate this variability and the factors that mold the outcome of ant–plant interactions are examined. In general, the majority of ant–plant mutualisms are facultative and involve groups or guilds of interacting species that vary in time and space, and that often affect each other by diffuse or indirect pathways. This discussion shows that obligate *Acacia–Pseudomyrmex* interactions, although well known, are unrepresentative of ant–plant mutualisms as a whole. The relative contributions of selection on ants and on plants become the basis of the final discussion on the issue "Are ant–plant mutualisms coevolved?"

There are many reasons why ants might provide excellent services for plants. Ants are abundant and diverse in most places where terrestrial plants grow. They are active twenty-four hours a day, either as individual species or as overlapping guilds of species foraging for particular periods during the day or night. Many fruit-, seed-, and pollen-dispersal agents are active only during the day and there can be intense competition among plants for their services. From this point of view it is perhaps particularly curious that night-blooming, ant-pollinated plants have apparently never evolved. Ants are commonly omnivorous, constantly switching between plant and animal foods. Plants can offer rewards that encourage ants to capture prey attacking their tissues and, as can be seen in Chapter 6, the rewards themselves may mimic prey. Colonies often respond quickly to high densities of food items, recruiting by means of pheromones or laying trails to major resources. This type of density-dependent response may be particularly beneficial to plants suffering infestations of enemies, or plants in the process of releasing large numbers of seeds. Also, the impact of ant foragers on plant enemies can be far greater than the actual number of prey they capture. The aggressive movements of ant foragers can dislodge some, frighten away others, and still more can be harassed during feeding, egg laying, courtship, or molting so that their impact on the plant is significantly reduced either directly or indirectly.

Ant colonies are often large and perennial so that plants at particular patches or in given habitats may benefit from sustained services. Greenslade and Halliday (1983) reported a colony of *Iridomyrmex purpureus* in South Australia that consisted of 85 nests, over 1500 entrances, and a territory of about 10 hectares. Perenniality in many kinds of colonies results from buffers against adverse conditions such as mechanisms of food storage. Some have adult castes known as repletes that sequester food in their gasters, which can become enormously distended. The larvae of others contain storage tissues and food may be obtained from them by inducing regurgitation, or by cannibalism. With the return of more favorable conditions starved colonies rapidly produce foragers, which often have an urgent demand for the nutrients in plant rewards. Many ant species nest in plants they service, and in addition to the advantage of proximity of ant foragers for defense, ammonia from decomposing nest wastes may be a valuable source of nitrogen for some plants or plant tissues.

Other potentially favorable aspects of colony organization and ant behaviors are discussed at length in subsequent chapters, especially Chapter 2. Among them, social organization itself, the liquid food requirements of ants, and differentiation into castes with specialized behavioral repertoires are the most important.

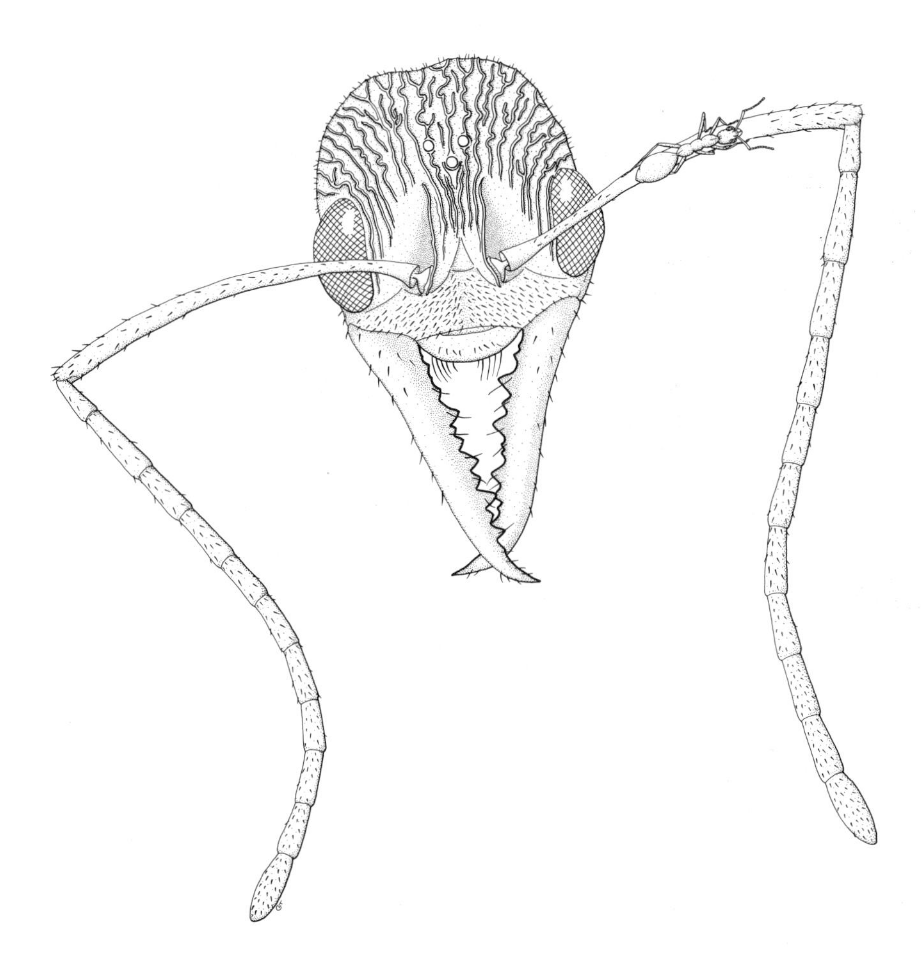

Figure 1. Size differences in ants: The head of *Myrmecia nigriscapa* bearing *Tapinoma melanocephalum*. Although these two species do not naturally occur together, size differences of this magnitude can routinely be found among ants foraging on a plant.

One aspect of the organization of ant communities produces more equivocal results from the point of view of plants. Ant rewards tend to attract a great variety of ants. In fact, plants themselves are popular foraging grounds for many kinds of ants irrespective of the presence of rewards. Figure 1 illustrates size differences in ant species that may be attracted to a single extrafloral-nectary-bearing plant species. This difference is the same order of magnitude as that between a deer mouse and a mountain lion, a human and a blue whale, or a sunflower and a spruce tree. Variation in the ants visiting rewards has potential advantages and disadvantages for the plants. On the positive side, an array of visitors is more likely to buffer the mutualism against discontinuities in the presence or activities of individual species. Furthermore, services such as defense may be not merely more continuous, but also more effective as a variety of ants will attack a variety of plant enemies. On the other hand, non-mutualistic ant visitors could sabotage interactions. For example, tiny ant species may not be able to move elaiosome-bearing seeds and simply gnaw the reward, in effect stealing it. Similarly, tiny ants taking extra-floral nectar may be quite unable to deter herbivores larger than themselves. At the other end of the scale, large aggressive and dominant ants that take rewards but provide no services can actively exclude ants more useful to the plants.

Evolutionary dilemmas and dialectics such as these are the stuff of much of evolutionary ecology, and ant–plant mutualisms provide a focus for the study of the ecological and evolutionary dynamics of simple and complex species interactions. The results of selection are often difficult to predict and present an array of interactions from those that appear hopelessly casual to others where the loss of one partner means the extinction of the other. Amid the almost baffling variation in function and effect, ant–plant mutualisms nevertheless reveal patterns if not laws, and probabilities if not certainties.

Ant–plant mutualisms are important also for their effects on community structure and organization. Ants protecting plants may affect the fecundity of their host on one hand, and the fecundities of a variety of animals on the other. Among these may be, for example, the larvae of pollinators, seed predators or herbivores of other plant species, or the juvenile stages of predators and parasitoids. The effects thus ripple through the community at various trophic levels. Other effects include the abiotic aspects of community organization such as nutrient cycling. The spring flora of many eastern U.S. forests, although spectacular while in flower, appears insignificant in terms of biomass and its impact on forest dynamics. However, these species take up nutrients such as potassium and nitrogen from the spring snowmelt – nutrients that would otherwise be lost to streams and rivers. The retention of nutrients by the lowly herbaceous layer is so important to forest nutrient cycling that it has been called the "vernal dam" (Bormann & Likens 1979). The point is that up to 35% of all herbaceous species, and up to 76% of all herbaceous stems, can be ant-dispersed plants. The plants are often organized into guilds, usually consisting of about seven species, the number determined in part by the availability of seed-dispersing ants. In fact, ant activity can be a good predictor of the number of ant-dispersed plant species at any given site in some forests, and a fair predictor of the number of all herbaceous species (Beattie & Culver 1981). Ant dispersal of vernal dam species exemplifies the potential influence of ant–plant mutualisms in community structure and organization.

The ramifications or ripple effects of ant–plant mutualisms are not always clear, but generally penetrate several trophic levels. A group of trees in a forest may be cleared of a defoliating pest by an ant species nesting among its roots. The ants may navigate according to the light patterns produced by the canopy (Holldobler 1980), and some of the plants forming the canopy may have grown from seeds planted by ants in their nests. Ants are also important tillers of the soil in many habitats. These diverse community effects begin with the dynamics of the ant–plant interactions themselves, and this is what this book is all about.

2

Origins and early evolution of ant–plant mutualisms

In this chapter I first examine fossil ants and plants and primitive living ants and angiosperms to try to reconstruct the origin and early evolution of ant–plant interactions. The fossil record is fragmentary and even extant primitive species may be only remotely related to those that were involved in the first interactions. This discussion is therefore speculative.

Ants and plants in the Cretaceous

It is now generally agreed that the flowering plants (angiosperms) had spread across most of the land masses of the world and had diversified dramatically during the early part of the Cretaceous period. By mid-Cretaceous, about 100 million years before present, this plant group was dominant among terrestrial vegetation (Raven 1977; Doyle 1978). The many suggestions for the causes of this comparatively rapid ascent include changing physical, climatic, and geographical conditions (Axelrod 1970), the rise of major insect pollinator groups (Takhtajan 1969; Crepet 1979), the appearance of avian and mammalian seed-dispersal agents (Regal 1977), the proliferation of herbivores (Ehrlich & Raven 1964; Burger 1981), and the evolution of novel plant secondary compounds (Swain 1977, 1978). In this chapter, I argue that another factor, the ants, contributed significantly to the success and adaptive radiation of the flowering plants.

Given that the fossil record places the angiosperm rise to dominance in the early to mid-Cretaceous, it is appropriate to ask when the ants began to flourish. The order Hymenoptera, to which the ants belong, first appeared in the early Triassic, perhaps 100 million years before the angiosperm accession. However, although the mid-Cretaceous is thought to be the period when the aculeate Hymenoptera underwent a first radiation (Crowson et al. 1967), the first fossil ant, *Sphecomyrma freyi,* was discovered in upper Cretaceous amber, and is therefore from a slightly later period (Wilson, Carpenter, & Brown 1967). Extensive paleontological collections from Siberia and Mongolia suggest that ants originated in the mid-Cretaceous, possibly in the Abian or Cenomanian periods, becoming abundant in the late Cretaceous (Professor V. V. Zherichin, personal communication). Brown (1973) summarized evidence that the Dolichoderinae dominated the early Tertiary, but was displaced from this position by a variety of genera in the Myrmicinae by the late Tertiary. However, as both subfamilies contain genera that interact mutualistically with plants, little can be made of this interesting change. The data permit us to speculate with some justification that the ants had evolved and undergone extensive adaptive radiation by the mid-Cretaceous, much as the angiosperms had done (Smart & Hughes 1972; Carpenter 1977; Doyle 1977).

Thanks to the work of many paleobotanists (summarized by Takhtajan 1969; Stebbins 1974; Doyle 1978) much is known about the anatomy and morphology of Cretaceous plants and the level of complexity to which they had evolved. In many early Cretaceous deposits angiosperms appear with a mixture of more primitive forms such as cycads, ginkgos, conifers, and ferns. Unfortunately there seems to be no fossil evidence for adaptations to ants by plants of this antiquity. This may be primarily because nobody has thought to look for it. There are, however, adaptations to ants in living preangiosperm groups, especially the ferns. Extrafloral nectaries are found in the fern genera *Pteridium, Platycerium, Polypodium, Drynaria, Angiopteris, Cyathea,* and *Hemitelia* (Lloyd 1901; Koptur, Smith, & Baker 1982), the last two being tree ferns with massive structures reminiscent of those of the Cretaceous era. The rhizomes of several other fern genera such as *Lecanopteris, Solanopteris, Polypodium,* and *Phymatodes* are riddled with tunnels that function as domatia for ants. Since plants may absorb nutrients from waste materials abandoned in the tunnels (Ridley 1910; Gomez 1974; Janzen 1974c; Huxley 1980), this is another example of mutualism between ferns and ants. In a few cases, the sori or clusters of sporangia bear tissues attractive to ants (elaiosomes) and are dispersed by them (Janzen 1974c). We can only speculate as to when these adaptations arose, but it is possible that ant–fern interactions are of Cretaceous age.

Two general statements can be made that lead to the conclusion that early ant–plant interactions involved a variety of Cretaceous plant groups and not just angiosperms. First, the characteristics of angiosperms that most definitively set them apart from their predecessors are predominantly associated with reproduction, notably the structure and antomy of the flower (Stebbins 1974). The important corollary to this is that most aspects of the vegetative structure of the angiosperms, although greatly advanced and refined by them, were present as primitive versions in preangiosperm groups. Second, in reviewing the entire range of ant–angiosperm interactions it is clear that the majority involved nonfloral structures. Adaptations to ants are primarily vegetative, involving roots, stems, and leaves (Wilson 1971; Beattie 1982). Consequently, it is these structures, and the level to which they had developed by the Cretaceous period, that may provide most insights into the beginnings of ant–plant interactions.

Large leaves (megaphylls), vascular tissues, and large branching structures were all present in one preangiosperm group or another. Large leaves with a net of veins are as old as the Permian Glossopterids and can be seen in ferns such as *Dipteris* and the gymnosperm *Gnetum* (Melville 1969). However, the development and structure of the angiosperm leaf are different from these, exhibiting for example, such advanced and novel features as intercalary growth and a hierarchy of successively finer orders of venation. These, in turn, have permitted the expansion of the leaf into a large laminar structure, supported by veins of various diameters and ramified throughout, from margin to midrib, by vascular tissue (Doyle & Hickey 1974). The appearance of this structure initiated the diversification of the angiosperm leaf into a host of new shapes, sizes, and types of venation (Hickey & Wolf 1975). This was probably associated with changes in overall size and height of plants. For example, taller species intercepting direct sunlight were likely to evolve "multi-layer light-gathering strategies" in which lobation, dissection, or pinnatification of leaves would permit light to filter through to lower layers (Horn 1971).

A parallel story of origin and elaboration applies to vascular tissues. Vessel elements are present in some ferns, horsetails, and gymnosperms but absent in a few genera of primitive angiosperms. However, in the angiosperms diversification and specialization of both xylem and phloem elements are far greater than in any other group (Stebbins 1974; Dickison 1975). New developments in xylem structure probably fed back positively to the evolution of laminar leaves (Takhtajan 1969).

Various other advances, especially in root anatomy, the protection of buds, the form of the stele, and its branching at nodes, together with leaf and vascular evolution produced a burst of vegetative diversification (Delevoryas 1962). Takhtajan (1969) discussed the evolutionary potential of the vegetative advances of the angiosperms. He pointed out that this group exhibits such great plasticity and variety of growth form that many niches can be simultaneously exploited in any given area to produce a multilayered community. In this setting more primitive groups such as gymnosperms and ferns were doomed to play only a minor role. This entire argument is best expressed by Doyle and Hickey (1974): "We suggest that definitive replacement of the conifers by angiosperms as forest-canopy dominants, made possible by the accumulation of sufficient vegetative advantages (e.g. vessels, leaf morphology) to outrival the conifers, did not occur until well into the upper Cretaceous."

The appearance of increasingly complex vegetative structures and communities was an essential ingredient in the evolution of insect–plant and ant–plant interactions. Vascular tissues would have provided a liquid food source for insects with sucking mouthparts. Laminar leaves with broad surface areas would have provided a tissue food source for insects with chewing mouthparts. The Hemiptera, especially Homoptera, with their plant-sucking mouthparts, were clearly abundant by the end of the Permian (Carpenter 1977). This method of feeding may have originated in the preceding Carboniferous period in the extinct orders Palaeodictyoptera and Megasecoptera (Scott 1977). By the Cretaceous the Hemiptera were very well established. Fossil evidence of herbivory by insects turns up as leaves that show insect damage (Plumstead 1963; Kevan, Chaloner, & Savile 1975; Scott 1977), coprolite containing leaf cuticle, epidermis, and other leaf parts (Harris 1964; Baxendale 1979), and damaged stems (Scott 1977). There seems to be a fairly clear distinction in the fossil record between these remains and those left by animal groups that fed on plant detritus. Many plant-chewing insect orders were abundant by the late Cretaceous, especially the Orthoptera, Coleoptera, Phasmatodea, Diptera, and Lepidoptera (Carpenter 1977). Some phytophagous invertebrates may have been huge; for example, *Arthropleura,* a strange primitive arthropod belonging to the extinct class Arthropleurides, grew up to six feet. Fossil remains show that its gut was packed with vascular tissue (Rolfe & Ingham 1967).

The impact of the ant's mode of feeding

This discussion of the fossil record leads to the following scenario: As the ants evolved, becoming more numerous and diverse, they encountered environments increasingly dominated by flowering plants. The vegetative structures were larger, more complex, and with the evolution of woody secondary thickening, more permanent. These vegetative structures were a novel array of potential resources, and yet they were to some extent already preempted by the more ancient groups of sucking and chewing invertebrates, especially other insects. There are many possible explanations as to how, in spite of this, the ants managed to join the crowd of herbivores exploiting the advanced Cretaceous plants and early angiosperm resources. One basic property of the ants, however, was probably fundamental to the process, namely, their mode of nutrition: Ants require liquid food.

All of the most primitive ants are carnivorous (Haskins & Haskins 1950; Brown 1954; Brown & Taylor 1970; Wilson 1971). (I reserve the term *predaceous* for the consumption of entire individuals, plants as well as animals. For example, granivores are seed predators.) Prey is subdued with powerful stings and large mandibles (Cavill, Robertson, & Whitfield 1964; Kugler 1980), and carried back to the nest where the integument is breached and fluids, especially hemolymph, sucked out. The infrabuccal chamber, located next to the mouth, filters out all but the minutest solid particles. For example, *Solenopsis invicta* filters out particles down to 0.88μ (Glancey et al. 1979). Furthermore, in some ant species, regurgitative feeding among workers results in the filtration of more and smaller particles. Eisner and Happ (1962) found that honey containing corundum particles of only 10 microns, when passed among *Camponotus pennsylvanicus* workers, was progressively diluted so that recipients at the end of a five-ant food chain had crops free of any particles. Following filtration, fluids are fully ingested and transferred to the crop via a long and narrow esophagus that passes through the slender pedicel to the gaster (Eisner 1957; Gotwald 1969).

Typically, in the primitive subfamilies Nothomyrmeciinae, Myrmeciinae, and Ponerinae, capture of prey involves some crushing with the mandibles together with vigorous stinging. Back at the nest gnawing and chewing with the mandibles (malaxation) often can be observed but serves only to gain access to the body fluids of the prey, and the fragments created by this crude mastication are not ingested. This mode of nutrition is apparently shared by both adults and larvae (Wigglesworth 1974). Indeed, larvae may be the principal individuals involved with digestion in some ant species, especially seed eaters. Even then, malaxation appears to be followed by larval extracellular digestion with the seed contents being liquefied prior to ingestion (Went, Wheeler, & Wheeler 1972; Abbott 1978; Davison 1982).

The carnivorous habit, although most obvious among primitive ants, has never been completely lost. The advanced subfamilies Myrmicinae and Formicinae exhibit a huge array of omnivorous genera that consume animal fluids when they are available. Even among leaf-cutting and granivorous ants insect prey constitutes a small part of the diet (Sheata & Kaschef 1971; Wilson 1971; Levieux & Diomande 1978). Occasional carnivory has been proposed as an adaptive mechanism to counteract low nitrogen availability in many plant tissues (Mattson 1980).

There are both morphological and physiological correlates to the carnivorous, liquid-food nutrition of ants. Chief among them are the absence of chewing mouthparts and cellulose-digesting enzymes. The mouthparts differ from those of typical plant-chewing insects in the following respects: The upper lip or labrum, which is a primary incisor in chewing insects, is poorly developed and concealed by the clypeus. The lacinia of the maxilla, which is strongly chitinized and armed with teeth in chewing insects, is soft and membranous in ants. The labium, also large and well developed in chewing insects, is small and less developed in ants (Wheeler 1910; Borror, Delong, & Triplehorn 1976).

There are many problems associated with the digestion of plant tissues, not least of which is the low protein and high cellulose content of many plant parts (Southwood 1972; Wigglesworth 1974). These problems are often exacerbated by the presence of toxic secondary compounds (Feeny 1976; Harborne 1978). The vast majority of cellulose ingested by herbivores is broken down by intestinal microorganisms whether the herbivores are invertebrates such as termites or vertebrates such as cattle. In fact, plant tissues often survive the alimentary canal almost intact, save where the chewing or grinding mechanisms have actually severed the cellulose cell wall and released the protoplasmic contents (Friend 1958). Very few animals secrete their own cellulose-splitting enzymes, and ants are not among this select group (Wigglesworth 1974).

Primitive carnivorous ants confined to a liquid diet would suffer some disadvantages but also would experience some advantages with respect to other invertebrate groups exploiting Cretaceous plants. On one hand, they would be unable to exploit the plant tissues directly, having neither masticatory mouthparts nor cellulose-digesting enzymes. On the other hand, if plant liquids were somehow available, they would not have to deal with the formidable cellulose barrier or, possibly, with defensive plant toxins discussed later. Most likely, plant liquids were available in two forms, highly processed to be sure, but in abundant supply. The first form was provided by the hemolymph of plant chewers falling prey to the ants. The fossil record shows that they would have included katydids, crickets, grasshoppers, cockroaches, beetles, stick insects, and butterfly, moth, and fly larvae (Carpenter 1977). The second form came from secretions from homopteran insects, especially families such as the Membracidae, Cicadellidae, Aphididae, and Coccidae. Today, these secretions (collectively called "honeydew") are often remarkably complete foods containing sugars, free amino acids, proteins, minerals, and some vitamins (Way 1963; Kennedy & Fosbrooke 1972). Some homopterans contain gut symbionts (microorganisms) that perform at least two functions of benefit to their hosts and consequently of potential benefit to the ants. First, they sometimes supply sterols, a major dietary requirement of many insects, including ants (Dadd & Krieger 1967). Second, although toxic secondary compounds such as alkaloids are frequently present in vascular tissue, both xylem and phloem (James 1950; Mothes 1960), gut floras may be capable of degrading them to harmless forms (Janzen 1979; and references therein). In addition, secondary compounds are often selectively distributed among plant tissues and aphids are adept at avoiding areas where they are concentrated (Mothes 1960; Janzen 1979). Perhaps Southwood (1972) best summarized the situation by calling the mixture of ingested fluids and microorganismal products in the gut a "brew" that "seems. . .well able to meet the nutritional requirements of the archaic terrestrial arthropod." This being so, ants would have found a very rich source of liquid food, comparatively free of the problems of digesting cellulose, nitrogen limitation, and toxic defensive compounds.

The transfer from carnivory to herbivory appears to be a very minor step. Ants are notoriously omnivorous. Even carnivores such as *Odontomachus* and *Paraponera,* with the most spectacular scythelike jaws, large stings, and aggressive behavior, harvest plant food (Horvitz & Beattie 1980; Young 1980, 1982). Very primitive carnivores such as the bull ants of Australia (genus *Myrmecia*) will gather immobile animals such as pupae, mobile animals such as caterpillars, and free-flowing plant exudates such as extrafloral nectar (Gray 1971; and personal observations). Unlike many invertebrate carnivores, ants do not require movement of the prey to trigger prey-capture behavior, so that animal and plant fluids are collected in many different forms, from living or moribund prey, fruits, nectaries, wounds in plants, and so on. Even the leaf-cutting ant, *Atta,* which was previously thought to feed exclusively on fungi, is now known to imbibe plant sap while preparing leaves for its "gardens" (Stradling 1978; Quinlan & Cherrett 1979). Thus, there seems no reason to postulate complex physiological, anatomical, or morphological changes in the transition to herbivory.

As the angiosperm evolved, new tissues and structures appeared. These were simply new surfaces to be explored by foraging workers. Much to the advantage of the ants they were already populated by herbivores converting cellulose-laden plant material of variable digestibility, nutritional value, and toxicity into more useful fluids. Perhaps at first the earliest carnivorous ants used the Cretaceous plants merely as expanded hunting grounds. We do not know how soon the secretions of plant suckers were exploited. Wheeler (1914) recorded aphids trapped with ants in Baltic amber, but these remains are from a much later period. However, an ability to occupy a variety of trophic levels at once - as carnivores, as herbivores taking nectar or sap, as secondary herbivores utilizing homopterans - appears to have been one of the keys to success.

Were the ants welcome on the Cretaceous plants and early angiosperms? Overall, the preponderance of adaptations to ants on vegetative structures are attractants, suggesting ants were of benefit. On the other hand, many features of reproductive parts (flowers and inflorescences) are interpreted as ant barriers and repellents. These range from physical barriers such as mucilaginous tissues and glandular hairs (Kerner 1878; Kevan & Baker 1983), repellent floral nectar or flower parts (Stager 1931; van der Pijl, 1954; Janzen 1977; Baker & Baker 1978; Stephenson 1981), or ant baits in the form of extrafloral nectaries (Bentley 1977a). Although these mechanisms are patently of limited success in many plant species, the ability to evolve ant repellents is clear. Therefore, it is reasonable to suggest that ants were not discouraged from foraging on vegetative parts.

Foraging of workers on plant structures probably reduced the numbers of some herbivores. Indeed, their very presence most likely diminished herbivore damage, at least on a local scale. Sap oozing from wounds caused by herbivores may have lured ants to the very places where their predatory behavior was most needed (Scott 1980). Through time some angiosperms evolved a variety of attractants and lures that had the effect of maintaining a cadre of ants on their structures; this in turn resulted in the attrition of herbivores. Vegetative structures were becoming larger, more complex, and more vulnerable to herbivores. If ants could obtain food from the plants, it would be advantageous to protect them in return.

The impact of social organization

What part did the social organization of the ants play in this process? The fossil record tells us little about the origins of sociality. There are two known fossils of Cretaceous age from true eusocial groups: *Sphecomyrma*, an ant discussed earlier, and *Cretatermes*, a termite. There is no information on the stage of development of their sociality, but *Cretatermes* is thought to have been fairly advanced (Wilson 1971; Carpenter 1977).

Assuming some degree of sociality among early ants it is possible to identify a few characters of social organization and angiosperm resources that might have generated positive feedback. The principal resource offered by angiosperms was probably food. However, most reviews on the subject do not place food high on the list of "prime movers" of social evolution (Wilson 1963, 1975; Hamilton 1972; Lin & Michener 1972; Alexander 1974). Food resources becoming available with angiosperms were most likely small individual items such as nectaries, herbivore prey items, and sucking insects secreting honeydew. In a large herb, shrub, or tree many hundreds or even thousands of such food items may be scattered among the roots, stems, branches, and foliage. Each item may be sufficient to fulfill, at least temporarily, the food requirements of an individual, or group of workers. However, back at the nest there are other adults and larvae to be fed, often in huge numbers. A colony consisting of many individual foragers is a solution to the problem of harvesting diffuse spatial and temporal patterns of small food items. One special feature of social organization in the ants appears to be of major selective advantage and that is differentiation into castes. A second feature, regurgitative feeding or trophallaxis, may have enhanced this advantage.

Caste differentiation in ants is accompanied by division of labor (polyethism) with the result that each one exhibits a limited repertoire of behaviors. Among the worker caste behaviors tend to be influenced either by physical or by age characteristics (Oster & Wilson 1978; Herbers 1979). Consequently, either by virtue of a particular size, morphology, or age, some workers become foragers, leaving the nest to locate and bring home food. Although foragers may exhibit a variety of behaviors ranging from following chemical trails to killing prey, the task is comparatively specialized, for example, being free of all reproductive and brood-tending chores. Oster and Wilson (1978) argue that three characteristics of food resources were major determinants of caste evolution: (1) the distribution of food items in space and time, (2) the size distribution of the items, and (3) the defenses of the food items. Among those ants considered to be phylogenetically the most primitive there is often little physical differentiation among workers (monomorphism) and castes are apparent only as some degree of behavioral specialization, which in turn may be only temporary. In the genus *Myrmecia,* for example, workers forage individually, usually taking prey items about the same size as themselves and often imbibing plant secretions on the way. This kind of foraging, in which there is little or no cooperation among the workers, is called "diffuse" foraging by Oster and Wilson (1978) and may have been the method exhibited by the first ants on angiosperms. Although little is known about the foraging of the most primitive living ants of the genus *Nothomyrmecia* it appears that it also is diffuse (R. W. Taylor, personal communication).

This basic form of foraging has been modified in at least two ways with the result that larger resources or more difficult prey items can be included in the range of accessible foods. First, a forager finding an item she cannot handle herself signals this to her nestmates, who then come to her aid. This is recruitment. Second, workers utilizing the same route to and from the nest produce a conspicuous and persistent trail, which may be kept clear of objects that impede the passage of foragers (Oster & Wilson 1978).

All three types of foraging can be observed on plants, for example, *Myrmecia* foragers combing shrubs individually, *Solenopsis* recruiting to a damaged fruit, or *Formica* clearing a "highway" to a favored tree full of aphids. The form, number, and degree of polyethism of foragers are determined by a wide range of constraints, not least of which is the resource-distribution function $R(t, a, s, y)$, which is the number of food items per unit of area at time t, of age a, size s, and nutritional value y, where age is the period elapsed since a food item became available to the nest, size refers to the dimensions and therefore the difficulty of harvesting, and the nutritional value is determined by the particular mix of nutritional, nonnutritional, and defensive compounds in the item (Oster & Wilson 1978).

Overall it is clear that given a scatter of food items that vary according to spatial and temporal distribution, recalcitrance, and nutritional content, the possession of a subset of individuals in a colony that specializes in their location and harvesting is of great selective advantage. This behavioral specialization, free of other essential (and potentially distracting) nest functions, is a major characteristic and consequence of the social organization of ants (Heinrich 1978). Caste differentiation would seem to be a basic requirement for a society with a resource distribution exemplified by the scatter of plant exudates, herbivore secretions, and herbivore prey that became available at the time of the appearance of the angiosperms. Suppose, for example, that the first ant–plant interactions involved ants similar to the extant genus, *Myrmecia*. This genus exhibits a variety of primitive anatomical, morphological, and behavioral characters, including some aspects of its social organization. For example, foragers do not cooperate or recruit. My own observations show that the food items they seek are scattered among the branches and foliage of plants. The activities of a single forager I watched will illustrate this. She ascended the trunk of a shrub until, at a height of about six feet, she ran along an almost horizontal branch to its tip. Once there she tried to obtain nectar from some flowers but they seemed too old. She turned back and reconnoitered several branchlets, found two active extrafloral nectaries, and paused to sip from each one. For the next twenty minues she rambled throughout the other branches of the shrub, pausing often but apparently finding nothing more to eat. Finally, at another aging inflorescence, she found a small caterpillar and quickly grabbed it with her mandibles. Within another five minutes she had returned via a maze of branches to the nest. This type of foraging, carried out by a large number of individuals, transforms a highly dispersed resource into a steady, nestward flow of food.

In addition to the advantage of foraging, caste differentiation also facilitates the defense of plant resources. It is commonplace for ants to defend territories against competitors and intruders with aggressive worker or soldier castes. The perenniality of the colonies of many ant species might have been an advantage to plants, creating a stable defensive system for long periods. Territories may coincide with discrete vegetative units such as trees, groups of trees, or tree species. In fact, trees, with their wealth of food items, are fairly defensible, although battles against intruders attempting to ascend the trunk or gain access from adjacent foliage may be frequent and the territorial boundaires constantly changing, at least in detail (Wheeler 1910; Bequaert 1922; Greenslade 1971; Majer 1976a; Holldobler & Wilson 1977; Leston 1978). Holldobler and Lumsden (1980) have embodied this in the concept of economic defensibility and have analyzed the defense of territories by *Oecophylla,* the weaver ant, which fiercely defends groups of adjacent trees. The food resources (nectar, homopterans, and prey) are renewable and scattered in trees that are fixed in place for many years. This relatively stable resource base is defended by distributing workers at the boundaries. It would appear that a three-dimensional territory would require large numbers of individuals to cover this area. However, the colony weaves leaves together all over the trees so that the nest is diffuse, and the food harvesters can also be boundary patrollers. Although this is probably a highly evolved strategy it illustrates the defensibility of plant structures by ants. Other defended vegetative units range from clusters of epiphytes or "ant gardens" (Wheeler 1921) to individual herbs or patches of vegetation (Brian 1956; Brian, Hibble, & Kelly 1966; Inouye & Taylor 1979). A small defensive effort may be all that is required to protect a large resource because of some aspects of plant form. A tree in an open forest, for example, may require only a defensive patrol on the lower part of the trunk and a few others where branches interlace with adjacent trees to minimize invasion by other ants and wingless herbivores. This would apply equally on the smaller scale of shrubs and herbs. Aggressive castes on one hand and plant form on the other may result in comparatively defensible, secure territories for the harvesting of food items. Early angiosperms may have been favored hunting grounds for this reason.

In ants the advantages of social organization and caste to the harvesting of plant-borne food items may have been enhanced by the mechanisms for handling and distributing liquid foods, i.e., trophallaxis. Unlike bees, ants do not store food in external structures such as comb cells or honey pots. Food is stored in the crop and, in many genera, distributed among other adults and larvae by regurgitation. Consequently, the crop has a social as well as an individual function. In fact, regurgitative feeding is a fundamental bonding process among the individuals of a colony. One often sees columns of slender foragers ascending a tree to feed on homopteran secretions or extrafloral nectar. The descent may be considerably less dignified, the gait reduced to a waddle, so distended is the gaster with fluids. In the nest, the ingluvial fluid is shared with other individuals.

Among the most primitive living ants nectar feeding is extremely common but trophallaxis is not. It has been observed in a few species of *Myrmecia,* for example, but it is not yet clear whether its absence is only apparent, being the result of insufficient detailed observations (Haskins & Haskins 1950; Haskins & Wheldon 1954; Freeland 1958; R. W. Taylor, personal communication). Trophallaxis becomes so important among some advanced ant genera that it is a part of the very fabric of society, and glandular secretions are discharged with the nutritive ingluvial fluid. It can only be hypothesized that fluids of plant origin, collected by specialized foragers and stored in their crops, were a positive selective force in the evolution of trophallaxis. With the ant system of foragers returning to the nest with individual loads, food sharing must have been essential for larval feeding. If a proportion of the food was already ingested as fluid, then trophallaxis among all ant colony members may have been a particularly easy step to take.

Early ant–plant habitats

Is there any evidence on the type of habitats in which ant–plant interactions would have been favored? Again this is a highly speculative subject and an intense debate continues, particularly on the origin of angiosperms. The prevailing view is that the angiosperms evolved in the tropics and subtropics, but there is much disagreement as to whether the first habitats were moist and equable or semiarid and seasonal. There is also deep disagreement on the growth form and structure of the first angiosperms, especially in relation to the flower (Stebbins 1974; Doyle 1978).

The classical theory of angiosperm origins proposes that the first angiosperms were large trees, similar to the extant Magnoliaceae, which evolved in warm, moist, equable tropical forests (Arber & Parkin 1907; Takhtajan 1969). An alternative hypothesis, championed by Stebbins (1974) and cautiously supported by others (e.g., Doyle 1978), suggests that the first angiosperms were more likely to have been shrubs growing in semiarid environments with limited and seasonal rainfall. This hypothesis has the strong, plausible argument that seasonal drought would constitute the most powerful selection pressure for the evolution of the distinctive structures characteristic of the angiosperms, especially the closed carpel, the extreme reduction of the gametophytes, and double fertilization (Stebbins 1974, Chap. 10). Whether or not this hypothesis survives the scrutiny brought about by fresh fossil evidence, it is interesting to note that although ants are diverse and abundant in modern tropical rain forests, their greatest diversity and abundance may be attained in semiarid environments dominated by shrubs that might well be considered the contemporary counterparts of Stebbins's early angiosperms (Bernard 1968; Brown & Taylor 1970; Pisarski 1978; Andersen 1982). For example, Anderson (1983) recorded eighty-six species from twenty-seven genera in two 50-by-25-m plots in the semiarid shrublands of northwestern Victoria. In fact, semiarid habitats are well known for the abundance of ant–plant interactions, a large number of the shrub species being ant dispersed, including the genus *Hibbertia* in the Dilleniaceae, which is considered to be very primitive (Berg 1975; Stebbins & Hoogland 1976; Milewski & Bond 1982; Westoby et al. 1982). Others bear extrafloral nectaries that are visited by ants (Hocking 1970; Janzen 1973b), and on a more subjective level one cannot help being impressed by the vast, almost incredible variety of ants foraging on and around the vegetation in these areas today. Stebbins (1974) pointed out that they also harbor a large proportion of the known examples of rapid speciation in plants. It is tempting to attribute this in part to ant–plant interactions since ants are such a major component of the plant environment.

3

Plant protection by direct interaction

Ants foraging on plants take a great variety of prey items including insects and other invertebrates that are either herbivores or seed predators. Therefore, the mere presence of hunting foragers can provide some defense against plant enemies. With few exceptions plants are hospitable foraging areas, and once ants have gained access, they will hunt and remove prey irrespective of the size, architecture, or morphology of the plant. Ants remove a great variety of animals that do damage even, as in the case of *Monomorium floricola,* entering the tunnels of leaf-mining beetles to kill the tiny occupants (Taylor 1937).

The protective character of ants foraging on plants has been recognized for hundreds of years. In various parts of China nests of the weaver ant *Oecophylla smaragdina* were taken from the forest around citrus and litchi groves and placed on branches of orchard trees. Branches close to the nest were smeared with wax to prevent the ants from leaving the trees, and until they established their food-gathering territory their diet was augmented with dog intestines or silkworm larvae. After several weeks the ants established territories and patrolled the trees aggressively for food. Bamboo poles were used to create bridges from trees with nests to those without, and the groves were soon a mosaic of *Oecophylla* territories. As long ago as the eleventh century A.D., the Chinese observed the ants removing a considerable variety of insect herbivores and seed predators in large numbers, including stinkbugs of the hemipteran family Pentatomidae, many of which feed on plant sap, and the larvae of the butterfly *Papilio demoleus,* which were killed by workers stretching the unfortunate victim in several different directions simultaneously and holding it in that position until it died. Other insects including stem borers were harvested, and it was apparently well known that trees without ants produced fewer fruits than those with ants and that the few that did form generally aborted before ripening (McCook 1882; Groff & Howard 1925). Ants continue to be used as biological control agents in pine plantations in various parts of China. For example, a single nest of

Polyrhachis dives is estimated to harvest about 2000 caterpillars of the pine defoliator *Dendrolimus punctatus* per day (Hsiao 1980).

In Europe various species of *Formica* have been utilized for biological control in "forest hygiene" programs for several centuries. For example, *F. polyctena,* which is a major predator of the green oak leaf roller (*Tortrix viridana*), reduces defoliation by this moth so successfully that it is used to control it (Gosswald 1951; Gosswald & Horstmann 1966). It has been estimated that a medium-sized nest of this ant with a territory of 0.27 hectares harvests approximately six million prey items in a year, many of them herbivores and seed predators (Horstmann 1972, 1974). Skinner and Whittaker (1981) showed that *F. rufa* significantly reduced the populations of lepidopteran larvae living in sycamore and oak trees in English woodlands.

A review of the feeding habits of *F. rufa* by Adlung (1966) presents many details of the massive numbers of insects harvested by this ant in European forests. An average nest appears to provide a high degree of protection against lepidopteran and hymenopteran (sawfly) larvae over an area of 1000–1900 m^2. Estimates of the numbers of insects harmful to plants brought back to nests averaged around four hundred thousand per year. Detailed studies of foraging by ants of the *F. rufa* group in the USSR have revealed that they remove a great variety of herbivores and seed predators from hawthorne (*Cretaegus* sp.) and oak (*Quercus* sp.), including snails, beetles, the larvae of flies, moths, butterflies and sawflies, and many pupae. Furthermore, more formidable or better protected prey such as weevils are driven off aggressively before doing any damage (Inozemtsev 1974). Some herbivore defenses are of little help against ants as shown, for example, by Apostolov and Likhovidov (1973), who watched ants tear open the silk cocoons of *Tortrix* moth pupae and imbibe the fluid contents on the spot. The prey was consequently carried back to the nest in liquid form in the crops of the foragers.

Inozemtsev (1974) estimated that four nests whose foragers covered an area of approximately 200 m^2 removed a minimum of 100,800 invertebrates from surrounding trees in twelve days. However, he also pointed out that the activities of these nests were modest compared to one studied by Strokov (1956), which captured 4500 sawfly larvae, 3500 pine beauty moth caterpillars, 7200 *Tortrix* caterpillars, and 6500 pupae in one twenty-four-hour period. Claims of even greater levels of slaughter and pillage can be found in several excellent reviews of research into biological control by ants in the USSR (Grimalsky 1960; Khalifman 1961; B. A. Smirnov 1962; V. I. Smirnov 1966).

F. aquilonia colonies are so effective against plagues of the geometrid moth *Oporinia autumnata* that "green islands" of intact trees, 40 m across, remain around the nests when the rest of the forest is badly defoliated (Figure 2; Laine & Niemela 1980). It is possible to reduce defoliation further by increasing the density of ant nests (Adlung 1966), although the most abundant tree species, the birch (*Betula* sp.), bears no special lures for ants.

Transplantation of ant nests demonstrates the willingness of ants to ascend and hunt in vegetation that is novel to them. These experiments serve to emphasize both the versatility of their predaceous behavior and the resulting protective nature of their activity, whether or not the plant species involved possess any adaptations to attract ants. For example, *F. lugubris* nests were transplanted from their native forests in northern Italy where the trees were chiefly spruce (*Picea*) with some larch (*Larix*) and beech (*Fagus*), to a region of Quebec dominated by red and white pines (*Pinus resinosa* and *P. strobus*) with a few spruce, balsam fir (*Abies*), and paper birch (*Betula*). Within a year the ants had settled in and were removing a wide variety of herbivores from the trees, especially Lepidoptera, Coleoptera, Diptera, and Hemiptera. During a single two-week period the ants were carrying an average of 2175 spruce budworm (*Choristoneura*) larvae back to their nests each day. In addition, the predatory activity of the ants significantly reduced defoliation in some of the tree species (Finnegan 1975; McNeil, Delisle, & Finnegan, 1977, 1978). A variety of other transplanted ant species, most notably *F. obscuripes* and *Dolichoderus taschenbergi*, have also successfully exploited unfamiliar vegetation as hunting grounds in northern forests (Bradley & Hinks 1968; Bradley 1972; Finnegan 1977).

In tropical forests many arboreal ant species are well known for the aggressiveness with which they hunt and remove herbivores. This behavior has been explored as a means of biological control in various crop species, especially cocoa and coconuts. Ants such as *Oecophylla longinoda*, *Tetramorium aculeatus*, and several species of *Crematogaster* effectively remove large numbers of orthopterans and hemipterans that damage host plants (Way 1953; Vanderplank 1960; Room 1971, 1973; Leston 1973; Majer 1976a, b; Taylor 1977). Soysa (1940), growing orchids commercially in Sri Lanka, placed specimens containing herbivorous larvae on trees occupied by *O. smaragdina*. Within twenty-four hours his plants were free of pests.

Protection by ants is extended to other smaller plants and vines. Risch and Carroll (1982) showed that *Solenopsis geminata* reduced herbivores

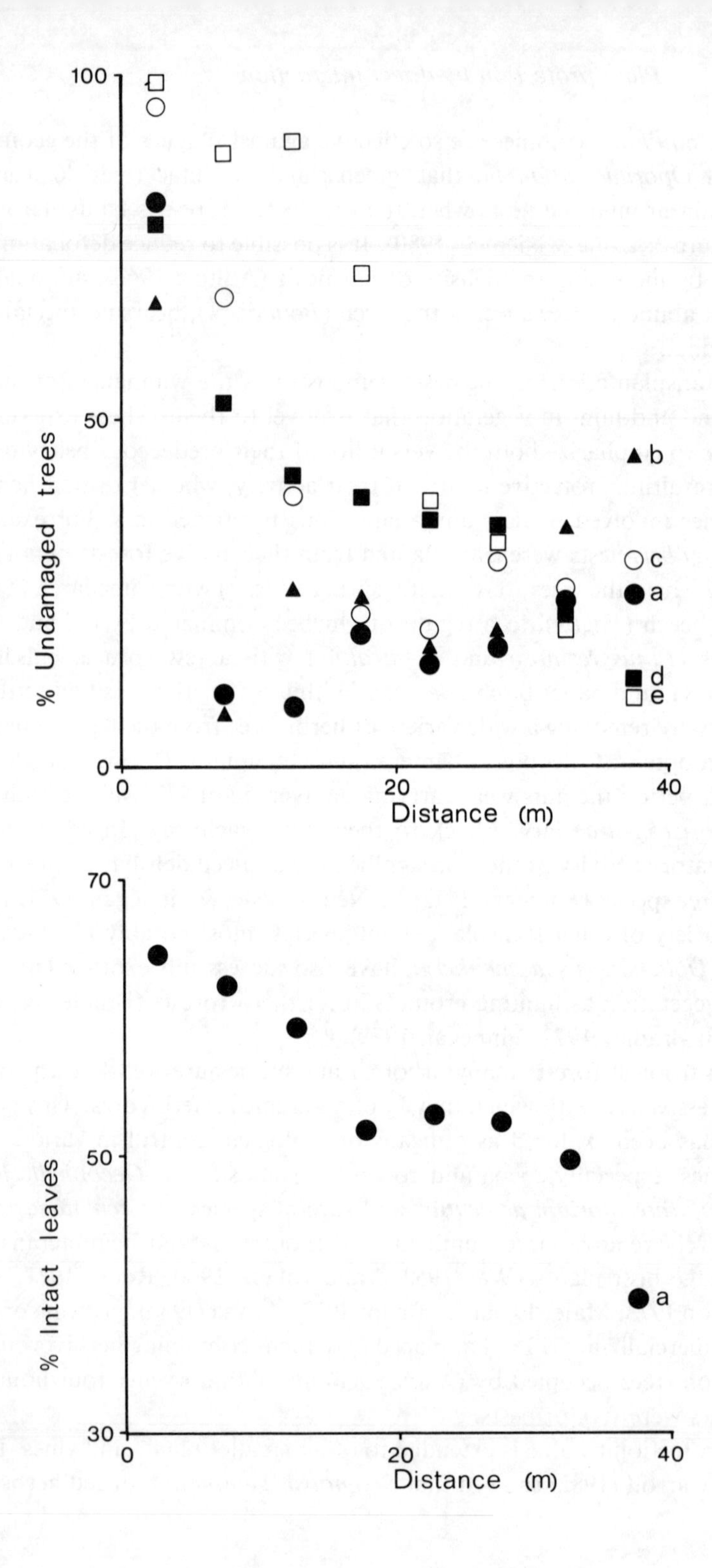

100
50
0
% Undamaged trees
0
20
40
Distance (m)
b
c
a
d
e
70
50
30
% Intact leaves
a
0
20
40
Distance (m)

on corn and squash in Tabasco, eastern Mexico. The lepidopteran larvae *Melittia* and *Diaphania,* which are particularly damaging herbivores, were reduced in numbers by as much as 82% and 73%, respectively, compared to plots where *Solenopsis* was excluded. The weevil *Sitophilus,* which attacks the seeds of corn, was reduced by 98% in the presence of this ant. A close relative *Solenopsis invicta,* which was accidentally introduced into the southern United States, has become a major predator of the boll weevil. Its impact on this seed predator has been sufficiently large to be characterized as "fortuitous biological insect suppression" by Sterling (1978), who suggested that ongoing investigations into biological control of the boll weevil may well have led to the deliberate introduction of the ant, had it not already spread to cotton-growing areas. That specialized adaptations are not required for ant protection is underlined by the studies of Pimentel (1955) and Pimentel and Uhler (1969), who recorded the wholesale destruction of dipteran eggs, larvae, and pupae by ants on garbage cans in Puerto Rico and chicken coops in the Philippines.

All of these studies illustrate that the mere presence of foraging ants on plants may reduce losses to herbivores and seedeaters. However, there may be a cost involved in this kind of protection that results from the omnivory of many ant species. Besides killing herbivores, they may cultivate homopterans for honeydew. This has been shown to decrease production in citrus groves (Sweetman 1958; DeBach & Huffaker 1971), coconut plantations (Way 1954), cornfields (Risch & Carroll 1982), and other agricultural settings where commercial criteria are applied. Indiscriminate predation by ants on insects beneficial to plants and plant enemies may result in a Volterra effect; that is, the ants reduce the numbers of the natural predators and parasites of plant enemies as well as the enemies themselves. Under these circumstances the population sizes of herbivores or seed predators can recover faster than those of their predators and parasites. Thus, indiscriminate predation by ants can lead eventually to serious damage to plants. This appears to apply particularly to agricultural and various man-made monocultural settings. Some workers in-

Figure 2 (*facing page*). Top: The percentage of undamaged birch trees at different distances from five wood ant (*Formica aquilonia*) mounds during an outbreak of the caterpillar *Oporinia autumnata* in northern Finland. The nests are labeled a–e ($r = 0.599$; $P = 0.001$). Bottom: The percentage of intact leaves on birch trees at increasing distances from the mound "a" during a year with normal herbivore density ($r = 0.92$; $P = 0.01$). (From Laine and Niemela 1980.)

vestigating the use of ants as biological control agents have concluded that they rarely protect plants because of the "homopteran problem." Although it may be true that ants make poor or erratic protectors in man-made vegetation, their impact in undisturbed environments may be more beneficial to plants. There have been very few studies of protection by ants in plant species that exhibit no obvious ant attractants and that are growing in pristine habitats. In these plant communities homopterans may not reach destructive densities, and costs to the plants may be either negligible or imperative in the interests of reducing herbivory. Evolutionary fitness, not bushels per acre, is the criterion of ant effectiveness in these circumstances. The ant–plant–homopteran interaction is complex and the outcome is variable (see Chapters 4 and 9). A second weakness of ant protection is that it may not be reliable. Many of the references cited present data showing that protection varies according to many factors such as availability of nest sites, shading, the appearance of alternative food sources, the health of the plant, and the outcome of battles with competing ant species. Some of the biological details are fascinating. For example, *O. longinoda,* which makes nests of leaves sewn together with silk exuded from larvae, may gradually abandon a favored territory as a result of infestation by pseudoscorpions and ant-mimicking spiders (Vanderplank 1960). As Bequaert (1922) noted, the notorious invasions of army or driver ants benefit plants: "Whoever has seen the almost fabulous numbers of individuals in the ant armies of the tropics can have no doubts as to the benefit they afford the vegetation by destroying caterpillars and other noxious insects." However, although many army ants such as *Eciton* sp. periodically bivouac for as long as three weeks (Schneirla 1971), what Spruce (1908) wrote is essentially correct as far as plants are concerned: "*Eciton* or foraging ants seem to be true wandering hordes, without a settled habitation." In other words, any plant protection they might provide is short lived and patchy.

There are four ways in which this basic ant–plant interaction has been modified with the result that foraging on plants has been intensified and protection rendered more effective. (1) Ants nest in a wide variety of plant structures most of which are abundant but fortuitous, such as dead twigs and abandoned insect galls. However, some structures are specialized and appear to have evolved specifically as nest sites for ants. (2) A few plant species produce specialized food bodies for ants on the leaves, petioles, or stems. (3) Extrafloral nectaries are present on many plant species. (4) Relationships with homopterans and lepidopterans may have evolved to such a degree that the losses of plant fluids may be outweighed by gains in ant protection. In general, these features localize ant activity on particular tissues and maintain it through periods of particular vulnerability.

Ant nests and domatia

Bequaert (1922) and Wheeler (1942) extensively reviewed the enormous variety of plant cavities utilized by ants as nest sites. Plant structures that could be reasonably interpreted as adaptations to facilitate nesting were called "domatia" by Wheeler. Hollow stems and twigs typically occur in many plant species, the cambium being active only at the periphery. Ants commonly nest in the hollow internodes and Wheeler (1942) listed many examples including *Bambusa* (Graminae), *Coccoloba* (Polygonaceae), *Terminalia* (Combretaceae), and *Xanthoxylon* (Rutaceae). In *Clibadium microcephalum* (Compositae) growing in Ecuador, the internodes are first hollowed out by the scolytid beetle *Scolytodes elongatus*. Several ant species are capable of taking over these cavities, especially *Pachycondyla unidentata* and *Pseudomyrmex* spp. (Nesom & Stuessy 1982). Ants commonly use cavities excavated by other animals. Stem and twig nesting becomes rare at higher latitudes and altitudes most likely because prolonged periods of cold either kill above ground plant structures or expose them to temperatures ants cannot tolerate.

The disused tunnels of tissue-boring insects and their larvae are utilized by many ant species. The tunnels may be in green living tissue such as those left by leaf miners, or in dead lignified tissues such as those left by wood-boring beetles and termites. Carpenter ants of the genus *Camponotus* may greatly extend the original nest cavity by gnawing away both living and lignified tissues. Many ants occupy galls, which are induced in plants by a host of organisms ranging from viruses to parasitic wasps. The hollow places resulting from the death and decay of trunks, stumps, branches, twigs, fruits, and seeds plus the irregularities of bark and wounds all provide nest sites. The natural cavities are often altered in shape and size by the addition of walls or plugs of soil particles, debris, or carton (chewed vegetable material, especially wood), which resembles coarse paper.

Many arboricolous ants of the tropics fiercely defend their nests. The weaver ant (*Oecophylla*) is a particularly well-known case because of its nests of freshly woven leaves and its ability to infiltrate a dozen angry workers inside your collar while you are preoccupied with the columns attacking your ankles. The enormous impact of such ants has been described by Carroll (1979), who compared the communities of stem-nesting

Table 1. *A selection of plant structures that bear ant domatia*

Plant structure	Example	Family
Leaf pouch or bladder	*Cola*	Sterculiaceae
	Cordia	Boraginaceae
	Tococa	Melastomaceae
Swollen petiole	*Tachigalia*	Leguminosae
	Piper	Piperaceae
Hollow stems	*Macaranga*	Euphorbiaceae
	Vitex	Verbenaceae
	Barteria	Passifloraceae
	Plectronia	Rubiaceae
	Cuviera	Rubiaceae
	Triplaris	Polygonaceae
	Cecropia	Moraceae
Hollow thorns	*Acacia*	Leguminosae
Hollow root	*Pachycentria*	Melastomaceae
Tubers derived from hypocotyl	*Myrmecodia*	Rubiaceae
	Hydnophytum	Rubiaceae

Source: Bequaert (1922), Bailey (1924), Wheeler (1942), van der Pijl (1955).

ants in Liberia with those inhabiting comparable forests in Costa Rica. The Liberian communities were dramatically depauperate compared to the neotropical ones. Carroll attributed this in large part to the bellicosity of a handful of West African arboricolous species that dominated the vegetation to the exclusion of other more docile ones.

Plant species that bear domatia are known as myrmecophytes or ant plants because of their intimate relationships with ants. Domatia are structures that encourage nest building by a subset of the ant species in the habitat. Table 1 gives examples of the structures and some genera that bear them.

The most famous domatia are the stipular thorns of some Central American and northern South American species in the genus *Acacia*. These have been the focus of a brilliant study by Janzen (1966, 1967, 1974a). The thorns are expanded and hollow (Figure 3). Ants of the genus *Pseudomyrmex* gnaw out a small entrance hole at the tip of one thorn of each pair, clear out the hypertrophying cavity, and move in. In addition to providing specialized nesting sites for ants, the acacias secrete nectar from large foliar nectaries and produce nutritive organs called Beltian bodies on the leaf pinnules. The Beltian bodies are rich in proteins,

Figure 3. Stipular thorns of a species of *Acacia* with domatia. The top thorn is cut away to show the cavity (in black) where the ants nest. An entrance hole is shown in the lower left thorn.

lipids, and carbohydrates and are eagerly harvested by foragers and fed to larvae. The nectar and Beltian bodies appear to provide a balanced diet for the ants. Janzen has shown that *Pseudomyrmex* colonies that take advantage of the food and housing provided by the tree exhibit extremely aggressive behavior toward almost any intruder. Although the ants are small they attack and repel large mammalian herbivores, destroy or drive off insect herbivores, and destructively gnaw invading vines and competing vegetation. In another genus *Cecropia* (family Moraceae), the domatia are hollow stems, the food bodies (known in this genus as Müllerian bodies) develop at the base of the petiole, and the ants belong to the genus *Azteca,* but their response to herbivores and vines is equally fierce (Janzen 1969a, b).

By growing myrmecophytes in greenhouses and gardens it has been shown that domatia are genetically determined and always develop, even in the absence of ants (Ridley 1910; Bequaert 1922). Careful anatomical and histological studies of domatia in petioles, stems, tubers, and thorns have shown that developmental hypertrophies of selected tissues give rise to the cavities in which the ants nest (Treub 1883; Bailey 1922a, b, 1923, 1924). There is no doubt that ants routinely accelerate the formation of the cavity by gnawing and removing the dying tissues, and may even extend it beyond its original boundaries. Ants gain access to the cavity by gnawing an entrance hole. This is of particular interest because ease of access is a fundamental characteristic of domatia and distinguishes them from a host of other types of plant cavities that remain inaccessible until breached by decay, herbivore damage, wounding, or fire. A section of the wall of the domatium is invariably thinner or less lignified than the rest, and soft enough to be tunneled by the ants. This accessibility may be short lived, appearing only when the structure bearing the domatium is young and unlignified. In *Cecropia* and *Macaranga* the points of access are depressions, just above the stem nodes, left by the axillary buds. The young tissues are sufficiently soft and compressed for the ants to chew their way through. Frequently the response of the plant to the creation of entrance holes is a lining of callus or sclerenchyma generated by the cambium, and if this response persists the ants are forced to maintain the holes periodically.

Perhaps the domatia best adapted to ant nesting may be defined, in part, by the quality of the entrance hole. A superior hole, for example, prohibits access by enemies but permits an internal circulation of air at the optimal temperature and humidity. Janzen (1974a) suggests that the holes on *Acacia* thorns do indeed accomplish this.

The quality of the protection afforded by ants inhabiting domatia and related structures is enormously variable. The most complete protection appears to accrue to the plant species that provides the greatest rewards. Thus, *Pseudomyrmex* on *Acacia* that is provided with year-round housing, nectar, and food bodies is an extremely aggressive ant that, in attacking invertebrate and vertebrate herbivores and competing plants, sets the standard for ant protection. Other *Pseudomyrmex* species on *Triplaris* and *Tachigalia* and *Azteca* species on *Cecropia* and *Cordia* appear to provide similar protection. *Pachysima* (*Tetraponera*) on *Barteria* chews invading vegetation and clears leaves of debris. Janzen (1972) identified several behavioral traits that suggest that *Pachysima* is especially effective against large mammalian herbivores, especially forest antelope and elephants. Most notably there is a "slow rain" of ants dropping off the tree so that a large mammal in its vicinity will probably be stung before it reaches the trunk. Second, the stinging behavior of the ants begins with a search for a vulnerable spot and climaxes with a prolonged sting that, in a human, produces deep-seated pain for one or two days. Third, *Pachysima* workers exude a noxious odor that is not an alarm reaction but may serve as a warning to large herbivores browsing at night. All of these traits contrast to some degree with those of *Pseudomyrmex,* which attack any foreign object as soon as it is encountered and quickly assemble a mass of workers that sting repeatedly no matter how small the invaders. Most insect herbivores on *Acacia* are quickly found and dispatched by *Pseudomyrmex,* but *Pachysima* may take several hours to perform the same service on *Barteria*.

About fifteen species of *Acacia* in East Africa bear swollen thorns. However, some individuals in any given population do not have any, which led Bequaert (1922) and Wheeler (1942) to postulate that the swellings were galls formed by dipteran or hymenopteran larvae and that once abandoned they made convenient nest sites for ants, especially of the genus *Crematogaster.* Hocking (1970, 1975) concluded that the galls were in fact ontogenetically determined and that the few individual trees without them were usually quickly destroyed by insect or ungulate herbivores. Hocking provided evidence that the ants residing in the galls afforded the plants some protection.

Thomas Belt (1874) was among the first to suspect the protective role of ants associated with domatia and he described how, whenever he touched the leaf pouches of various melastomaceous plants, "small black ants would rush out and scour the leaf in search of the aggressor." And yet protection by ants living in domatia is sometimes ineffective either because the herbivores have evolved defenses against them or because the "correct" ants are not in residence. An example of the first case is the *Acacia–Pseudomyrmex* interaction in which Janzen (1966) describes a beetle (*Pelidnota*) with an integument impenetrable to ants, a moth caterpillar (*Coxina*) that successfully fights back, and another moth larva (*Syssphinx*) that the ants simply ignore. In addition, the ants provide little protection for *Acacia* seed crops, most of which are destroyed by bruchid beetles (Janzen 1974a). Wheeler (1942) reported that *Azteca* on *Cecropia* did not deter sloths from consuming the foliage and that several lepidopteran larvae ate the plant, one of which (*Heliothis*) occasionally even resided with the ants in the domatia, at least while the colonies were young. Various beetle larvae bored into the stems and leaves, and one species of *Coclomera* was ignored by the ants although it caused heavy damage to the leaves. Beetles even consume the terminal meristem of new shoots (O. Taylor, personal communication). Bailey (1924) reproduced a photograph of a representative branch taken from a *Cordia* plant inhabited by the ant *Allomerus* in which the leaves had been shredded by the leaf-cutting ants (*Atta*). Perhaps the most extreme statement came from Wheeler (1942), who studied *Cordia alliodora,* which has internodal stem domatia, from Mexico through Central America, to Peru and Bolivia. He documented forty-four different ant species occupying domatia in this geographical area and yet declared, "No evidence was found to indicate that any of the ants attacked any of the leaf-eating arthropods." Needless to say, Wheeler was not a great believer in ant protection.

This quotation serves to emphasize a second reason why the protective system may break down: Either the ant species that exhibit the most protective behavior are not present, or they are present but in insufficient numbers to control the populations of herbivores and seed predators. Bailey (1923) described how the petiolar domatia of *Tachigalia paniculata* were generally occupied by beetles of the genera *Coccidotrophus* and *Eunausibius* while the plant was in its juvenile stages. The beetles drove away and even killed *Azteca* queens seeking domatia to start colonies. Sooner or later a queen succeeded in becoming established and the first generation began a war on the beetles, which invariably suffered defeat. It is interesting that this species of *Tachigalia* does not bear extrafloral nectaries or food bodies and that the protection appears to be of low quality. Protection also seems to be of low quality in species where the domatia themselves offer little protection to the ants, as in the case of leaf pouches (Bequaert 1922). This point will be taken up again in Chapter 5 as defense may not be the principal plant benefit derived from some domatia inhabitants.

As we have seen, variation in the degree of protection is evident even in obligate ant–plant interactions. For example, *A. melanocerus* is occupied by *P. satanica,* which is restricted to this plant species. It is particularly aggressive, as its name suggests, and occupied plants tend to be kept free of pests and vines. However, trees not colonized by this ant are colonized by other species of *Pseudomyrmex, Crematogaster, Azteca, Camponotus,* and *Paracryptocerus,* which perform very poorly in comparison to *P. satanica.* If the *Pseudomyrmex* is removed from *A. cornigera,* the tree invariably becomes so badly defoliated and overgrown by the surrounding vegetation that it dies within two years. A similar fate awaits trees not naturally colonized by the correct ant species, and it should be emphasized at this point that domatia can be occupied by a great variety of ants. For example, Wheeler (1942) listed thirty ant species inhabiting eight species of *Cecropia* (although many of them were only occasional), twelve ant species living in one species of *Triplaris,* and nine ant species in one species of *Tachigalia.* These data were augmented by similar lists from other domatia-bearing plants and by many observations on the variable effectiveness of different ant species with respect to plant protection. For example, among the six species of *Azteca* commonly inhabiting *Cecropia, A. alfari* was described as "the *Cecropia* ant par excellence." The occupation of domatia by spiders, dipteran larvae, beetles, and a variety of other nonmutualist invertebrates may eliminate protection by ants.

On reflection, the behavior of ants that results in the removal of vines, the clearing of competing vegetation rooted under the canopy and foliage overhanging it is remarkable. *Acacia, Cecropia, Barteria,* and other plant species with high-quality domatia and food rewards all benefit from this activity. Janzen (1969a, b) has called it "allelopathic" behavior since it can suppress competing plants with the same effectiveness as some allelopathic chemicals. Where this behavior originates is uncertain. Attacks on invertebrates, especially insects, are likely to be routine hunting behavior. Attacks on large vertebrates that cannot be food items can be interpreted as an extension of hunting behavior or protection of nest sites. It is probably elicited by similar stimuli, especially chemical signals, which excite the olfactory receptors of the ants. Attacks on plants might fit into this scheme if the ants obtain food such as sap from the damaged tissues. Otherwise their behavior may be nest cleaning similar to that exhibited by *Pogonomyrmex* and some *Formica,* which cut or gnaw vegetation encroaching on the nest.

It is of interest to note that even protection against encroaching vegetation may fluctuate in intensity. Keeler (1981b) showed that *P. belti* on *A. collinsii* will forage on the extrafloral nectaries of *Ipomoea carnea* vines that touch the tree. She did not report the outcome of this "infidelity" by the ants but suggested that it resulted from aging trees having insufficient food rewards for young and vigorous ant colonies. Perhaps for a while the vine augments the *Acacia* food supply and the nests continue to protect the tree. Alternatively, this may be the first step in the dissolution of the ant–plant interaction.

Food bodies

The term *food body* includes a wide variety of small epidermal structures that have been interpreted as adaptations to attract ant foragers. A difficulty with this interpretation is that epidermal structures in higher plants are so diverse in form and function that the subset to which food bodies belong is neither discrete nor uniform. Consequently, other interpretations are both possible and plausible. This difficulty is compounded by the fact that collection of food bodies by ants has been seen in relatively few plant species, and their actual consumption by adults or larvae is a still rarer observation.

Having said this, a few cases of food body production (Table 2) appear to be associated with ant protection or myrmecotrophy (see Chapter 5). The most complex food bodies are those of some of the same Central American *Acacia* species that bear domatia and are formed by additional meristematic growth at the tip of each rachis and pinnule of the compound leaf. It is differentiated anatomically into several tissues including a vascular bundle and a parenchymatous cortex of cells containing protein stored as tubules and lipids stored as droplets. Ants have been repeatedly observed harvesting these bodies and their larvae are known to eat them (Janzen 1966; Rickson 1969, 1975). Most species of the neotropical genus *Cecropia* produce food bodies that are avidly collected and eaten by ants. It is remarkable that the principle storage product is identical to animal glycogen, a molecule that is extremely rare in plants (Rickson 1971, 1973, 1976).

The Beccarian bodies of *Macaranga* belong to a very large category of structures called pearl bodies in which the predominant metabolite is lipid. They are generally distinguished from other epidermal structures such as trichomes and glands by a pearl-like luster, a small size (up to 3 mm), and the basal constriction, which makes them easily detached. In addition, as O'Dowd (1982) has shown, they are often associated with other ant attractants such as domatia and extrafloral nectaries. O'Dowd (1980) made a close study of the pearl bodies of *Ochroma pyramidale,* the balsa, a neotropical tree in the Bombacaceae. They are usually club-

Table 2. *Food body–ant associations in five plant genera*

	Acacia sp. (Leguminosae)	*Cecropia* sp. (Moraceae)	*Macaranga* sp. (Euphorbiaceae)	*Ochroma* (Bombacaceae)	*Piper* sp. (Piperaceae)
Major ant associates	*Pseudomyrmex*	*Azteca*	*Crematogaster*	*Solenopsis, Azteca*	*Pheidole*
Type of food body	Beltian, tip of pinnule and rachis	Müllerian, produced in large numbers on trichilium (a pad of tissue at base of petiole)	Beccarian, on stipules or young leaves	Pearl body, on leaves and stems of saplings	On petiole margins
Principle nutrient offered	Protein, lipid	Glycogen, protein	Lipid, starch, some protein	Lipid, perhaps some starch and protein	Lipid, protein
Anatomy	Multicellular, tissues differentiated	Multicellular	Multicellular	Multicellular	Single celled
Extrafloral nectaries present?	Yes	No	Yes	Yes	Yes
Domatia present?	Yes	Yes	Yes	No	Yes?
Plant provides complete diet for ants?	Yes	Together with coccid honeydew, probably yes	Together with coccid honeydew, probably yes	No	No
Do ants protect plants?	Yes	Probably	Probably	Probably	Probably

shaped, 0.5–3-mm long, and scattered on the leaf veins, petioles, and stems in such numbers that first-year seedlings bear a mean of 3157 ± 974 bodies. They average 74% (dry weight) lipid, in energetic terms about 0.2% of the energy devoted to leaf tissues. In assays of ant response to pearl bodies workers of *Iridomyrmex, Rhytidoponera,* and *Chelaner* collected pearl bodies and returned them to their nests. Although pearl bodies are reported to occur in fifty genera scattered in nineteen families, collection by ants has been observed in only six cases. Much work needs to be done on them, especially their role, if any, in ant-guard systems.

Risch and Rickson (1981) showed that the unicellular food bodies of *Piper cenocladum* are induced by the presence of the ant *Pheidole bicornis.* The curled margins of the leaf petiole produce a nest site that, when occupied by ants, produces about 1500 food cells, but when empty manages only about 50. Because the *Pheidole* remove small herbivores and invading vines the interaction is clearly mutualistic. Rewards, especially in the form of food, are commonplace in mutualistic interactions and are frequently vulnerable to theft by nonmutualists. The *Piper–Pheidole* case is the first in which the theft of rewards is minimized by their production only when both mutualists are present.

When food bodies occur in association either with domatia or extrafloral nectaries, or both, ant protection appears to be relatively effective. Most of the plants in Table 1 are more frequent in secondary growth, light gaps, creek banks, tree falls, landslides, human clearances, and the like, where both herbivore activity and competition from neighboring vegetation are especially fierce. Yet most of the studies of food-body-bearing species discussed earlier produce evidence that their presence induces sufficient ant protection to keep herbivore damage low and invading vines and other plant competitors at bay. Being relatively free of such burdens these species may rapidly overgrow their neighbors so that they are unshaded. Downhower (1975), for example, showed that ant guards of *Cecropia* are most abundant on leaves in the upper parts of the tree. Thus, young leaves are kept especially free of herbivores and encroaching plants. Downhower believed that this contributed significantly to the rate of vertical growth and hence to the success of the tree in maintaining its canopy in full sun, above its competitors.

One factor contributing to the effectiveness of food bodies is that when they are present with other ant attractants the ants may receive a more nutritionally complete diet from the host plant. A high quality of ant protection is correlated with this. Carbohydrates are often supplied by extrafloral nectaries or by homopteran honeydew, and the food bodies usually contain significant quantities of proteins and/or lipids. The lipids are probably a second major factor rendering the food bodies so potent in attracting and maintaining ant guards. Lipids play many essential roles in insect metabolism, behavior, and reproduction as attractants, arrestants, pheromones, and oviposition cues (Dethier 1947; Gilbert 1967). Others, notably sterols, are structural components of insect membranes and precursors of juvenile hormones and ecdysones (Chippendale 1972). Consequently, lipids are much sought after by insects, especially when they are dietary requirements (Dadd 1973). Several plant-feeding insect species are known to convert plant sterol into cholesterol, which is then metabolized into compounds indispensable to normal growth and reproduction. When placed in this context it comes as no surprise that ants eagerly gather lipid-containing food bodies. Certainly lipids are potent ant attractants and this will be discussed at length in Chapter 8. One indication of the importance of food bodies is the observation by Janzen (1965) that *P. fulvescens,* which is an obligate mutualist with *A. cornigera,* does not lay eggs in the absence of food bodies.

Extrafloral nectaries

The term *extrafloral* in fact refers to nectaries that are uninvolved in the pollination mechanism and is a misnomer since they can occur on the flower. A good rule of thumb is that nectaries occurring within the perianth are most likely to be pollinator attractants, whereas those outside the perianth, including its abaxial surfaces, are usually not (Baker, Opler, & Baker 1978). Extrafloral nectaries have been reviewed by Bentley (1977a) and Elias (1983) and are commonly found on leaves, stems, petioles, stipules, and the stalks and bracts of inflorescences or flowers; they are less commonly found on the abaxial surfaces of sepals and petals.

Extrafloral nectaries are very much more abundant than food bodies, being found in at least sixty-eight families (Elias 1983). They vary greatly in anatomy and morphology. The simplest are small groups of cells that appear indistinguishable from their neighbors except that they secrete nectar. Macroscopically they are often revealed only by the presence of nectar droplets or a cluster of ants feeding at a particular spot. On the other hand, complex extrafloral nectaries are cavities opening to the exterior by a slot or pore, and filled with glandular trichomes. At this level of organization the nectary may be vascularized by xylem and phloem and large enough to be conspicuous to the naked eye (Figure 4).

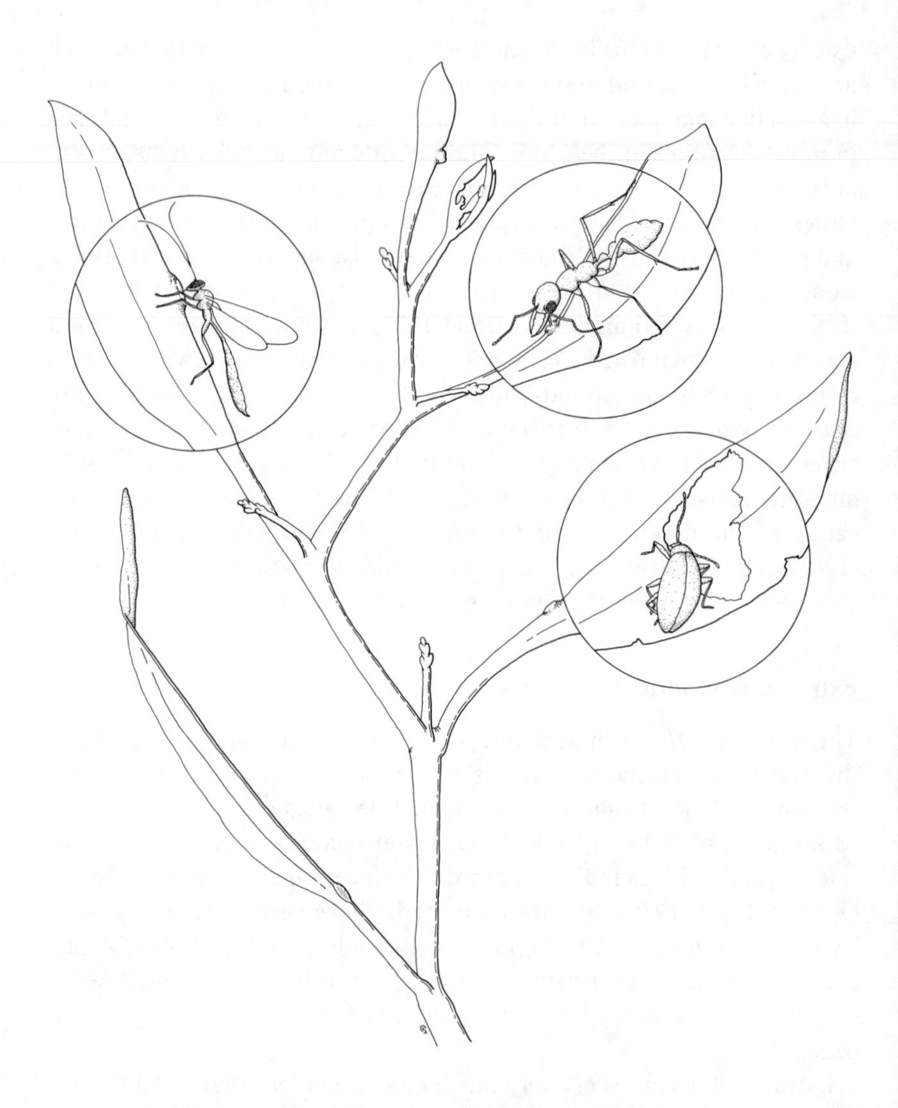

Figure 4. Extrafloral nectaries on the young phyllodes of a new shoot of *Acacia pycnantha*. Note the large size of the nectary on the youngest phyllode at the top. The insects shown represent a common situation in nature with the nectaries being visited by an ant (*Myrmecia* sp.) and an ichneumonid wasp, while a small beetle continues to feed.

The adaptive significance of extrafloral nectaries has been a bone of contention for decades. On one hand it has been argued that they are excretory sites where, for physiological reasons, fluids exude at the surface. Frey–Wyssling (1955), for example, envisaged the nectary as a site where

unwanted components of phloem strands were eliminated. On
hand, a huge variety of animals have been observed feeding
floral nectar, and it has been hypothesized that some of them
plant from harmful forms of herbivory and seed predation.
monly feed at extrafloral nectaries, treating them as integral pa[illegible]
territories, and behaving aggressively toward other visitors including herbivores (Bentley 1977a). Supporters of the physiological view ("exploitationists") claim that the plants do not benefit from the attraction of feeding animals to the nectaries, but that this form of herbivory is benign as only waste exudates are imbibed. By contrast, "protectionists" claim that the benefits that accrue to the plant secreting extrafloral nectar are of adaptive significance in reducing herbivore and predator damage. Plant fitness is increased by selection for traits that favor ant visitation to extrafloral nectaries.

Some extrafloral nectaries may well have started as epidermal sites that were physiologically superactive. The simpler ones may be little more than this. Others may have evolved following the expression of mutations subject to selection that gave rise to sites of exudation or secretion. Because the fluid balance of plants is critical to several fundamental physiological processes, a "leak" of this nature would at first appear to be a serious disadvantage. However, ants may have been sufficiently ubiquitous and abundant to counter such deleterious effects and act as a very powerful selection pressure.

For many years the rivalry between the hypotheses went unresolved largely because there was little evidence to provide a link between ants feeding at the extrafloral nectaries and ants actually reducing herbivory and predation. Indeed, several authors lamented that they never actually saw ants visit extrafloral nectaries or witnessed ants attacking herbivores. However, as long ago as 1889 von Wettstein had shown that the flowering heads of two species in the Compositae that bear extrafloral nectaries, *Jurinea mollis* and *Serratula lycopifolia,* were tended by ants. His exclusion experiments showed that the presence of ants reduced the damage to seeds caused by beetles and hemipteran bugs. In recent years several studies have greatly expanded von Wettstein's early evidence for the protectionist hypothesis, some of which are summarized in Table 3. The plants in Table 3 include a temperate vine (*Ipomoea*), a high-altitude herbaceous perennial (*Helianthella*), a tropical shrub (*Costus*), and warm-temperate trees (*Catalpa* and *Acacia*). Additional data can be found in Bentley (1977b), Tilman (1978), Pickett and Clark (1979), and O'Dowd (1979). Many studies have shown that the secretion of extrafloral nectar is greatest during periods of rapid vegetative growth such as the expansion

Table 3. *Some effects of ants on plants bearing extrafloral nectaries*

Type of damage to plant	Plant species	Plants without ants	Plants with ants	Source
Destruction of stigmas by grasshoppers	*Ipomoea leptophylla*	74% ($n = 138$)	48% ($n = 380$)	Keeler (1980)
Seed destruction by bruchid beetles	*Ipomoea leptophylla*	34% ($n = 149$)	24% ($n = 3509$)	Keeler (1980)
Mean number of seeds per plant	*Ipomoea leptophylla*	45.2 ($n = 271$)	403.2 ($n = 2419$)	Keeler (1980)
Number of sitings of parasitic flies on inflorescences (wet season)	*Costus woodsonii*	872	128	Schemske (1980)
Mean number of seeds per inflorescence (wet season)	*Costus woodsonii*	183	612	Schemske (1980)
Mean number of insect predators per capitulum	*Helianthella quinquenervis*	7.6	2.9	Inouye & Taylor (1979)
Percent seed predation	*Helianthella quinquenervis*	43.5	27.6	Inouye & Taylor, 1979
Number of mature fruits per branch	*Catalpa speciosa*	0.85 ± 0.81	1.11 ± 0.81*	Stephenson (1982b)
Average number of psyllid nymphs per shoot	*Acacia pycnantha*	351 ($n = 69$)	145 ($n = 72$)	O'Dowd (pers. comm.)
Shoot tips destroyed by psyllid nymphs	*Acacia pycnantha*	34% ($n = 67$)	7% ($n = 69$)	O'Dowd (pers. comm.)

* $P = 0.025$. All others are significantly different at $P = 0.001$.

of leaves and that the presence of ants is highly correlated with these peaks of nectar flow (e.g., O'Dowd 1979; Pickett & Clark 1979). This has suggested that ants reduce the losses of new shoots to herbivores and that a major benefit to the plant is continued vegetative expansion, especially vertically, so that a competitive advantage over neighbors is maintained.

Other benefits may also result from the reduction in losses of metabolites to herbivores and from concomitant gains in metabolites allocated to reproduction. Several authors have suggested that reduced seed set in the absence of ants may result in part from increased herbivory of leaves, rather than direct predation of seeds. Stephenson (1982b) found significantly fewer fruits maturing on unpatrolled branches of *Catalpa* trees and showed that mean leaf area on branches with ants was 817.3 ± 294.5 cm^2 compared to 563.9 ± 212.9 cm^2 on branches without ants – a statistically significant difference. O'Dowd (1979) showed that leaf damage in *Ochroma* saplings was 0.1%–5.45% of total leaf area when ants were present and 0.1%–23.07% when they were absent. This was also a significant difference although seed set could not be measured on trees this young.

The direct protection of flowers and fruits by ants attracted to extrafloral nectaries associated with these structures has also been reported (see Figure 5; Inouye & Taylor 1979). Keeler (1977) attributed reduced levels in the theft of floral nectar in *Ipomoea carnea* to the activities of ants visiting extrafloral nectaries on the flower stalks. Von Wettstein (1889) and Kerner and Oliver (1894) reported that ants protected the capitula of several genera in the Compositae with nectaries on the involucral bracts, spraying intruders with streams of formic acid. Heads of *Cirsium discolor* produce a very sticky secretion that traps harmful insects, and Willson, Anderson, and Thomas (1983) have produced some evidence that seed set can be enhanced through this phenomenon alone. *Mentzelia nuda* postfloral nectar is secreted at the top of the ovary for about ten days after the completion of anthesis. Ants visit the developing fruit (a capsule) during this period and their activities increase seed set significantly (Keeler 1981a).

Some authors have actually observed ants taking herbivores or seed predators. Stephenson (1982b), for example, observed 3858 ants descending the trunks of *Catalpa,* a tree with extrafloral nectaries, on the return journey to their nests. Twenty-nine of them carried an egg or larva of the principle herbivore, the sphinx moth *Ceratomia catalpae.* Tilman (1978) observed ants attacking and removing tent caterpillars on cherry trees. This predatory behavior was confined to the period, shortly after bud-

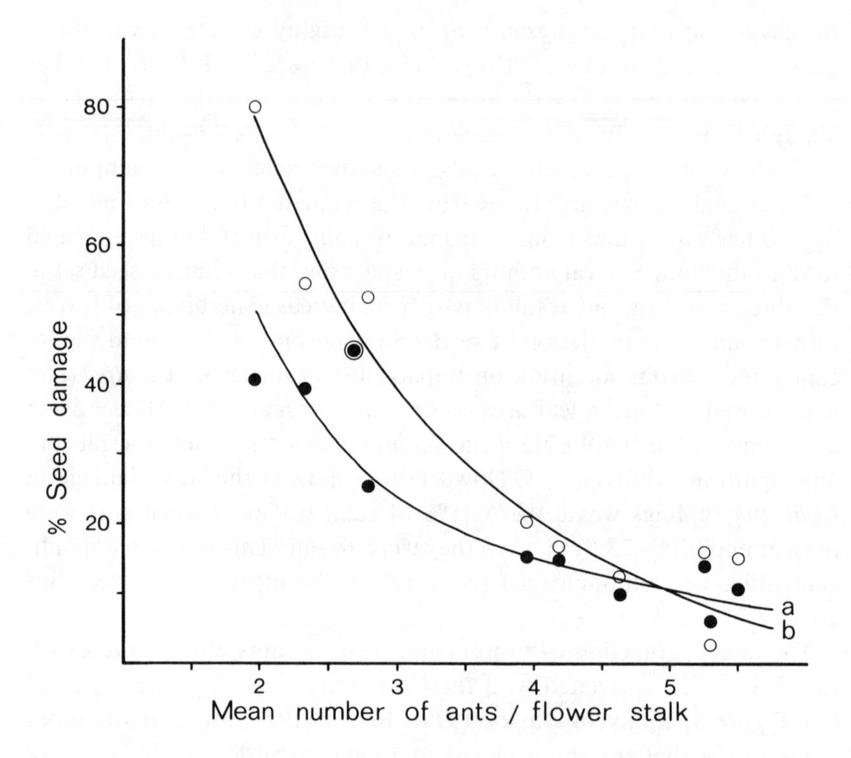

Figure 5. Relationship between the mean number of ants observed per flower stalk of *Helianthella quinquenervis* during the day and the degree of seed damage by dipteran and lepidopteran predators. The open circles (b) represent terminal flowers ($r = -0.923$; $P = 0.01$); the closed circles (a) represent axillary and terminal flowers together ($r = -0.976$; $P = 0.01$). Ants appear to be very effective at deterring seed predators. (From Inouye and Taylor 1979.)

break, when the availability of extrafloral nectar was at its peak. He also noticed that when ants attacked beetles, the victims would simply let go and fall to the ground. Bentley (1977a) pointed out that damage to plants would be reduced not only when ants carried herbivores back to the nest, but also when they dislodged them from their feeding positions. This includes many kinds of small herbivores that can eat vast amounts if left undisturbed but that may perish once dislodged, either because they cannot find their way back to the food, or because they are picked up by cursorial predators, including other ant species.

Aside from the foregoing examples, relatively little is known about precisely which plant enemies ants actually deter. The standard for ant protection has been *Pseudomyrmex,* the highly visible, aggressive, and numerous guard of *Acacia* that is capable of deterring enemies as diverse as vines, lepidopteran larvae, cattle, and field ecologists. Several invertebrate enemies have in fact breached this system, and even ant-guarded *Acacia* trees suffer herbivore damage and great losses of seeds. Clearly protection does not necessarily involve easily observable, nasty ants whose multiple stings linger vividly in the memory of the observer. Protection may more usually result from diffuse or even cryptic ant behaviors. Examples of diffuse behaviors include the general patrolling of plants, which results in the occasional capture of prey items, dislodging herbivores such as beetles, which fall to the ground, or frightening ovipositing lepidopterans and dipterans, which are forced to fly away. Observations of *Myrmecia pilosula,* an Australian bull ant that leaps at its prey, revealed its ability to inhibit an unknown dipteran's oviposition on *Acacia* by running energetically after the fly and jumping at it whenever it attempted to land. More cryptic behavior has been investigated by Letourneau (1983), who placed termite eggs on the leaves of species of *Piper* that provide domatia for the ant *Pheidole.* Although termites do not eat living plant tissue, their eggs are similar in size and shape to those of many herbivores. In a series of experiments the ants located and removed approximately 75% of all eggs, always within an hour. Curiously, few were eaten, the majority simply being dropped to the ground. This study highlighted the probability that the removal of insect eggs or early larval stages, which involves unspectacular or cryptic behavior, is as crucial to plant defense as a mass attack on a passing vertebrate herbivore.

Bentley (1977a) presented a series of ant characteristics that would contribute to effective protection of plants bearing extrafloral nectaries: (1) aggressive predatory behavior whereby any other animal is investigated and/or attacked; (2) twenty-four-hour activity as much herbivory occurs at night; (3) ability to nest in plant cavities as the predatory behavior would be reinforced by territorial defense; and (4) large numbers of foragers. In turn, the plant should provide a reliable flow of nutritious nectar, particularly during periods when its organs are most vulnerable, and in the immediate vicinity of those organs.

Extrafloral nectaries generally attract a variety of ant species as can be seen in Table 4 and Figure 6. Little specialization on the part of either the ants or the plants is evident from the data, although a single ant species may predominate in a particular habitat or on an individual plant (O'Dowd 1979). Schemske (1982) has shown that among species of *Costus* in Panama the array of ants on extrafloral nectaries varied according to

Figure 6. Two ants feeding at an extrafloral nectary on a phyllode of *Acacia*. The large ant is *Myrmecia pilosula* and the small ant is a species of *Iridomyrmex*. The outcome of encounters between these two ants is not easy to predict as sometimes the larger will drive away the smaller, but on other occasions a *Myrmecia* will retreat from an *Iridomyrmex*.

Table 4. *Ant assemblages at plants with extrafloral nectar*

Plant	Ants		Locality	Source
	No. genera	No. species		
Helianthella quinquenervis	3	5	Colorado	Inouye & Taylor (1979)
Ipomoea pandurata	7	10	N. Carolina	Beckman & Stucky (1981)
Pteridium aquilinum	7	8	New Jersey	Tempel (pers. comm.)
Mentzelia nuda	4	5	Nebraska	Keeler (1981a)
Costus woodsonii	4	7	Panama	Schemske (1980a)
C. pulverulentus	9	12	Panama	Schemske (1982)
C. scaber	9	13	Panama	Schemske (1982)
C. allenii	11	17	Panama	Schemske (1982)
C. laevis	12	24	Panama	Schemske (1982)
I. carnea	6	10	Costa Rica	Keeler (1977)
Calathea ovandensis	9	16	Mexico	Horvitz (pers. comm.)
Bixa orellana	8	12	Costa Rica	Bentley (1977a)
Aphelandra deppeana	ND	8	Costa Rica	Deuth (1977)
Ferocactus gracilis	10	11	Mexico	Blom & Clark (1980)
Acacia pycnantha	13	24	S. Australia	O'Dowd (pers. comm.)
Helichrysum viscosum	9	13	S. Australia	O'Dowd & Catchpole (1983)

Source: Adapted from Schemske (1983).

plant species and according to the height of the inflorescence above the ground. The latter effect was due to the addition of arboreal ant species in the taller plants. The behavior of different species of ant visitors varies greatly as Horvitz and Schemske (1984) have shown (Table 5). This study has clearly demonstrated the consequence that ant species differ in their effectiveness as protectors against herbivores and seed predators.

Table 5. *Quality of defense by particular ant species*

Ant species	Body size (mm)	Number of inflorescences	Seed production/inflorescence*		Ant activity/ inflorescence (mean)***
			Mean**	SD	
Wasmannia auropunctata (Roger)	1.0	35	28.0A	18.69	0.67
Crematogaster sumichrasti Mayr	3.5	21	19.2AB	8.33	0.65
Solenopsis geminata (Fabr.)	3.5	14	16.0BC	9.24	0.70
Brachymyrmex musculus Forel	1.5	127	15.8BC	12.98	0.66
Monacis bispinosus (Oliv.)	5.0	11	15.5BC	12.96	0.51
Paratrechina spp.	2.0	15	15.1BC	10.31	0.58
Pachycondyla unidentata Mayr	7.0	17	12.1BC	10.21	0.43
Pheidole gouldi Forel	4.0	16	10.2C	9.43	0.57

* Untransformed means and SD are reported, but significance testing was performed on transformed data.
** Means followed by the same letter are not significantly different ($P < 0.05$, Tukey's Test).
*** Ant activity is defined as the proportion of census dates for which ants were observed on a given inflorescence.
Source: Horvitz and Schemski (1984).

Several studies of extrafloral nectaries have differed from most of those cited so far in that the authors have found (1) evidence of protection in some habitats but not others (Barton 1983); (2) equivocal evidence of ant protection (Longino 1983); and (3) no evidence at all of protection (Boecklen 1983; O'Dowd & Catchpole 1983; Tempel 1983). In two Australian *Helichrysum* species that bear extrafloral nectaries on the floral capitula, the exclusion of ants led to an increase in the number of other insects visiting the flowers and the fruits. However, it did not lead to an increase in the number of adult or immature seed predators on or in the capitula, and no significant differences in seed set were found between ant-excluded and control capitula (O'Dowd & Catchpole 1983). Studies such as these that do not support the protectionist hypothesis need not send us scurrying back to the exploitationists' fold. Variation in the function and effects of interactions such as these is to be expected, and an examination of their dynamics is likely to be as fruitful in discovering the role of extrafloral nectaries in plant protection as searching for alternative hypotheses (see Chapter 9). However, in reviewing the literature it has been clear that some claims for ant protection have been made on excessively slender evidence. Careful experimental design is necessary for the demonstration of ant protection that results in an increase in some component of plant fitness. Also there is evidence for other functions for extrafloral nectaries, as the remainder of this section shows.

Extrafloral nectaries may be the basis for another form of plant protection whose importance probably has been greatly underestimated. In addition to ants, many of the organisms that feed on extrafloral nectar are predators or parasites of herbivores. More than a century ago Trelease (1879) proposed that by attracting predators and parasites to extrafloral nectaries, plants might reduce levels of herbivory and seed predation. It is not clear which group predominates among these visitors since few systematic studies have been done, but wasps, including both predators and parasites of herbivores, are clearly of major importance. Larval nutrition in the majority of wasps is carnivorous. The larvae of predatory wasps are fed freshly killed and malaxated prey. In the huge group of parasitoid wasps, larvae generally emerge from an egg laid on a host that they gradually consume. Most adult predatory or parasitoid wasps feed on floral or extrafloral nectar, and the plants to which they are attracted are often their principal hunting grounds (Evans & Eberhard 1970; Spradbery 1973).

Wasps are abundant and diverse and their attraction to extrafloral nectaries may constitute a danger to herbivores as great as ants, to which they are closely related. The problem of observing an ant actually disturbing or killing a herbivore, discussed earlier, is often compounded when observing predatory or parasitic wasps by their fast, agile flight, small size, or cryptic habits. However, some data support the idea that wasp guards are important in reducing herbivory. In his study of extrafloral nectaries of *Catalpa speciosa,* Stephenson (1982b) noticed the pupae of the parasitoid wasp *Apanteles congregatus* "on the exteriors of hundreds of *Catalpa* sphinx moth larvae." Bequaert (1922) listed various wasps as visitors to extrafloral nectar, and Hetschko (1908) listed three species of predatory vespid wasps, ten species of parasitic ichneumonids, and two coccinellid beetle species visiting the extrafloral nectaries of a single plant species, *Vicia sativa,* in central Europe. Another legume, *Calpurnia intrusa,* attracts a great variety of wasps in South Africa including seventeen species of Mutillidae, and Jacot-Guillarmod (1951) especially noted various ichneumonids, evaniids, and braconids attracted to the foliage and combing the leaves and stems for victims. In extensive studies of neotropical rain-forest butterflies, Gilbert (1975, 1980) and Gilbert and Smiley (1978) have emphasized the combined impact of ants and parasitoid wasps, which, having been attracted to extrafloral nectaries on the larval food plants of the butterfly *Heliconius,* significantly reduce the numbers of caterpillars.

Putman (1963), observing the extrafloral nectaries on peach trees, recorded that they were attractive to several species of parasitoid wasps. These observations were made in an orchard and wasps have been investigated, and utilized, for the biological control of crop herbivores and seed predators for a very long time (Sweetman 1958; Swan 1964). Several thousand parasitoid wasp species have been described and many more await discovery. The array of those potentially available for biological control is staggering (Huffaker 1971). In the protection of both wild and domesticated plants the specialization of many genera and even subfamilies plays an important role. For example, various genera of the Braconidae and Ichneumonidae specialize on foliage or wood borers such as moth larvae, beetle larvae, or sawfly larvae. Other genera in the Chalcidae and Encyrtidae specialize on particular groups of Homoptera. It is probable that all invertebrate herbivores are subject to attack by one kind of wasp or another (Borror et al. 1976).

The role of wasps, especially the Vespidae, as predators of herbivores is well established. Many other kinds of predators may also be involved and although this discussion is straying from the central theme of protection by ants, the research into the effects of extrafloral nectaries in cotton is worth mentioning. These glands attract a variety of insects, including the ants that attack major predators such as the boll weevil (Bentley 1983). Yokoyama (1978) also showed that predaceous anthocorids and lygaeids feed on both extrafloral nectar and phytophagous thrips. The bugs do not control the thrips but apparently eat them in large numbers. The role of other predators in extrafloral nectary defense systems is largely unknown.

The problems of demonstrating wasp-guard extrafloral nectary systems in the field are great. Individual predatory events are often extremely difficult to witness, or to predict. The cues utilized by parasitoid wasps to locate their prey are often extremely subtle. For example, some braconids home in on compounds volatizing from the leaves of the food plants upon which their prey is feeding, whereas others respond to compounds emanating from the dung of the victim (Vinson 1976). Exclusion experiments that demonstrate the effects of ant guards are difficult to devise when flying insects of varying sizes and seeking such subtle visual or olfactory cues are involved. This is undoubtedly why this aspect of extrafloral nectary ecology has been neglected to date. There are additional complicating factors. First, many other types of predators are attracted to extrafloral nectaries that might constitute a significant part of plant defense. For example, Putman (1963) and Stephenson (1982b) observed coccinellid beetles visiting the nectaries and then feeding on homopterans. Second, both extrafloral and floral nectar may be involved in defense against herbivores. Many parasitoid and predatory wasps feed at flowers but are unlikely to carry out pollination (Muller 1883; Knuth 1906–9; Proctor & Yeo 1973). Their visits may be adaptive, nevertheless, in encouraging them to hunt in the vicinity of the nectar source. Visitors that take floral rewards but do not pollinate are widely regarded as "thieves," but in cases such as these the plant may benefit from a reduction in herbivory. Third, the reverse may be true; that is, extrafloral nectaries are part of the pollination system. Nectar-seeking pollinators may be initially attracted to them but then move on to the flower.

An example of extrafloral nectar serving as a pollinator attractant is *Acacia terminalis.* Floral nectaries are absent but large extrafloral nectaries (2 mm–12 mm in length) are present on the petiole of each leaf and secrete large quantities of nectar during flowering. The extrafloral nectaries attract ants, bees, many other kinds of insects including parasitoid wasps, and several kinds of birds. Knox et al. (in press) have shown that the nectar is rich in hexoses and contains as many as eighteen amino acids. This mix of sugars and amino acids is generally considered to be particularly attractive to birds (Baker & Baker 1975). The nectaries themselves

: frequently a red or red brown – colors attractive to birds (Faegri & ı der Pijl 1966). When birds such as the grey-breasted silvereyes (*Zos- ops lateralis*) and the eastern spinebill (*Acanthorhynchus tenuirostris*) feed on the nectar, their heads and shoulders are dusted with pollen from the globular inflorescences strategically positioned close to the nectary.

In summary, extrafloral nectaries may play various roles according to the species and location of the plants that produce them. They may form the basis of a relatively simple ant-guard system in some situations, or the core of a complex ant-guard, wasp-guard, pollinator-attractant system in others. Much work remains to be done in this field.

A closer look at ant guards and extrafloral nectaries

Given the wide range of plant growth forms that bear extrafloral nectaries, and the variety of habitats in which they occur, it seems likely that the particular properties of ant protection are of selective advantage to certain plant structures and tissues, rather than particular species in particular habitats. Thus, all plant species must expand new meristems and leaves at intervals that vary according to life history, intensity of competition, seasonality, and so on. These tissues are generally the most vulnerable to enemies, being soft and chewable, highly palatable, generally rich in the limiting nutrient nitrogen, and sometimes low in toxins (Orians & Janzen 1974; Rosenthal & Janzen 1979; Mattson 1980). However, their maturation leads to important changes defensively. As meristems and leaves cease to expand they become tougher, more fibrous, or lignified, and they may acquire prickles or secondary toxic compounds (Feeny 1976). Therefore, during periods of growth a potentially large number of maturing structures are exposed to short periods of acute vulnerability. This situation may be exacerbated by growing in disturbed habitats where the rapid and continuous production of new shoots is at a premium to outcompete vigorous, neighboring secondary growth.

Bentley (1977a) suggested that many species that produced extrafloral nectar are plants of disturbed habitats and secondary growth. This seems to be true of some of the ant-guarded trees, for example, *Acacia* (in Central America and Australia), *Cecropia, Triplaris, Macaranga,* and *Ochroma* (Spruce 1908; Holttum 1954a, b; Willis 1966; Janzen 1974b) and various shrubby and herbaceous species. However, the same review by Bentley lists many species that are not characteristic of these environments and that occupy more stable habitats either in the tropics or the temperate regions. Therefore, the claim that ant-guard systems are characteristic of early successional, fast-growing, often weedy species of ephemeral habitats is only partially true. In fact, any growth form, from annual to herbaceous perennial or climax forest tree, faces the same basic problem of shoot production at some time or another. Furthermore, some types of defenses are inappropriate for expanding leaves and meristems; for example, they cannot become lignified or they would cease to expand. Toxic secondary compounds, which would appear to be an appropriate defense for these tissues, vary widely in concentration according to species (Feeny 1976; Coley 1983), although high concentrations of defenses such as simple and condensed phenols have been found in the young leaves of a variety of tropical trees (Milton 1979; Oates, Waterman, & Choo 1980; Coley 1983). Even when toxins are present in young structures the question remains as to the particular selective advantage of extrafloral nectaries.

The answer appears to be found in the characteristics of both the defended tissues and the properties of ant defense. Expanding shoots are not only vulnerable, but they are also particularly nutritious. Coley's (1983) analysis of the young leaves of tropical trees has shown that they are higher in nitrogen and protein than older leaves. This seems to be fairly general (Mattson 1980), and hence young shoots are the focus of attack for a great variety of herbivores, both invertebrate and vertebrate (e.g., Milton 1979; McClure 1980). The combination of high vulnerability and high nutritive content of young shoots is a crucial weakness in the vegetative demography of plant populations. Thus, it is reasonable to suppose that selection may result in the evolution of a variety of defense mechanisms, one of which is geared specifically to young growing shoots.

The importance of ant protection of young shoots and leaves may be related to the disadvantages of other kinds of defenses. To begin with, expanding organs cannot be "ossified" by the addition of fiber, cellulose, or lignin – common defensive compounds that render tissues too tough or unpalatable to eat. Second, the defensive strategy of unpredictability, or escape in time and space, can be relatively ineffective, at least for young plant parts (Coley 1980). This is probably because large numbers of vulnerable parts appear all at the same time, thus offering a bonanza to herbivores even if the plant is somewhat isolated. In other words, new growth can be both predictable and apparent (Feeny 1976). Third, there seem to be severe limits on the type and concentration of chemical defenses that can be sequestered in young shoots and leaves (Coley 1983, and references therein). These limits may result from the energetic expense and adverse physiological consequences of loading them with repellent toxins. Such "costs" as these are poorly understood. However, the energetic and metabolic costs of ant defense may be low; O'Dowd (1979), for example, showed that extrafloral nectar production in *Ochroma* was only 1% of the energy invested in leaves. Also, there may be further "savings" as the nectaries themselves are often anatomically and morphologically simple, requiring little differentiation or additional structure in order to function. Coupled with this is the intimate association of many kinds of extrafloral nectaries with vascular tissue. Nectar production does not involve the synthesis of secondary metabolites and so proximity to vascular strands containing primary metabolites under pressure may render its secretion relatively cheap. On the other hand, some evidence suggests that ant defense may be too expensive, too inefficient, or both to maintain over the long term. Studies have repeatedly shown that extrafloral nectar production is limited to periods of shoot and leaf expansion. As soon as expansion has ceased and the organs reach mature size, other more permanent defenses such as lignification or digestibility-reducing substances take over. Janzen (1973a) suggested that food-body production is expensive in *Cecropia* as it is lost when herbivore pressure is reduced (Chapter 9).

Ant defense has some particular strengths and weaknesses relative to other kinds. Chemical defenses are generally most effective against either specialists or generalists, not both, and their success may be limited to a particular group of enemies (Rhoades & Cates 1976). Ants, on the other hand, attack without regard to the chemical susceptibilities of their enemies. Specialists who have breached ant defenses do occur, but a chemical specialist remains a prey item to an ant and is likely to be removed regardless. In this context, the attraction of a variety of ants to extrafloral nectaries appears to be of great advantage. It has been shown in this chapter that this is generally true and that "assemblages" (Schemske 1982) are the norm. Obsevations of assemblages on Australian acacia bushes showed they may consist of only a few ant species, but the service they provide may be fairly comprehensive: A tiny but very numerous species of *Iridomyrmex* combs even the smallest cavities for prey, removing eggs and early instar larvae. A species of *Camponotus* or *Polyrhachis* patrols the bush for larger prey, while a handful of very large bull ants (*Myrmecia*) move nimbly from branch to branch taking large insects and threatening vertebrate intruders. Curiously, this mix of ant species rarely led to antagonism; a *Myrmecia* readily yielded extrafloral nectaries to *Iridomyrmex,* possibly because the latter exudes chemical repellents.

Since an array of ants hunts an array of prey, selection for a variety of ants offering protection may be expected.

Extrafloral nectaries also attract many other insects, especially other Hymenoptera such as parasitoid and predatory wasps. As some of these may constitute a form of plant protection, it seems likely that selection should favor freely available nectar, which contains a variety of rewards. In contrast to ants, wasps have the disadvantage of foraging only during the day. Although the species composition of ant assemblages on plants may change through the twenty-four-hour period, predator ants are present both day and night. In fact, the foraging of ants on plants at night is so routine that extrafloral nectaries may be an important mechanism to attract ants into the vegetation during the day when they are vulnerable to predation and dessication. Be this as it may, the importance of nocturnal defense by ants cannot be exaggerated as much herbivory and seed predation takes place under the cover of darkness. This alone suggests that extrafloral nectar is primarily a general ant reward, capable of maintaining foragers on the plant around the clock.

4

Plant protection by indirect interaction

Homopterans

Ants are notorious for the habit of maintaining colonies of homopterans on plants (Figure 7). The principal families involved are the Membracidae or treehoppers, the Cicadellidae (Jassidae) or leafhoppers, the Psyllidae or "lerps," the Fulgoridae or planthoppers, the Aphididae or plant lice, the Coccidae or soft scales, and the Pseudococcidae or mealybugs. Together they represent thousands of different interactions with ants, the majority still undescribed and unstudied. Interactions between homopterans and ants have been reviewed by Way (1963) and many are fundamentally mutualistic. The Homoptera secrete honeydew on which the ants feed. In return, the ants provide a number of vital services to the homopterans.

Homopterans take sap directly from the phloem through the slender mouthparts. Phloem contents are under several atmospheres of hydrostatic pressure and little effort is required for ingestion. However, the animals are capable of regulating their intake (Kennedy & Fosbrooke 1972). Changes in the chemical constitution of the sap occur during its passage through the homopteran gut so that when it becomes available to ants as honeydew it contains a variety of sugars, organic acids, alcohols, plant hormones, salts, vitamins, amino acids, and amides (Brian 1977). The main nitrogenous components of honeydew are amino acids, but these fluctuate widely according to the condition of the host plant (Mittler 1958; Llewellyn, Rashid, & Leckstein, 1974). However, a low nitrogen content in the sap stimulates the homopterans to increase the rate of feeding. The presence of ants elevates feeding rates so that they, in turn, may receive a volume sufficient for their own nitrogen demand (Way 1963; Kennedy & Fosbrooke 1972).

Homopterans mutualistically associated with ants produce a drop of honeydew at the anus when palpated by an attending worker. The ant then imbibes the drop in a process called trophobiosis. Some aphids possess a ring of setae around the anus that hold the drop in place while the

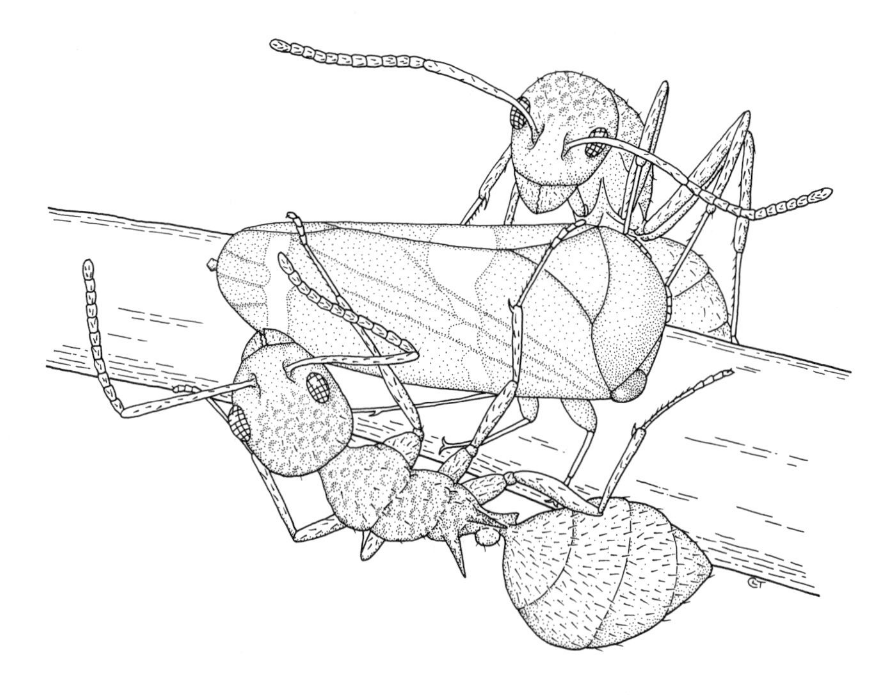

Figure 7. Two ants of the genus *Polyrachis* tending a cicadellid homopteran on a twig of *Eucalyptus*. The ants have been stroking the leafhopper with their antennae and the lower one is turning toward the anus anticipating a drop of honeydew.

ant feeds. In the absence of ants the honeydew is ejected from the anus in various ways. When the homopterans are at low densities the honeydew is merely scattered around adjacent stems and leaves, forming a sticky, sugary deposit. Fungi that develop on this substrate can reduce photosynthetic activity in leaves and may even stimulate abscission. However, this rarely appears to be a major problem for the plant (Banks 1962; Borror, Delong, & Triplehorn 1976). On the other hand, if the homopterans reach high densities in the absence of ants the slow rain of honeydew and its accumulation on their bodies result in potentially harmful fungal infections (Strickland 1947, 1951; Way 1954). Consequently, in some situations and for some homopteran species, the ants perform an important service in maintaining the hygiene of the homopteran colony (Beyer 1924; Majer 1982).

Most authors agree that the main service ants provide for homopterans is protection, both from enemies and the weather. Ants drive away or kill predators especially larval coccinellid beetles and syrphid flies, which can drastically reduce the numbers in unattended colonies (Banks 1962; Way 1963; Burns & Donley 1969; Bristow 1982). Also homopterans are hosts to a myriad of parasitoid wasps, but ants can reduce their activities dramatically, often by preventing gravid females from landing and ovipositing on the victim (Bartlett 1961). Ants also protect their homopteran colonies from enemies and rain by building shelters over them. These may be constructed out of soil particles, pieces of vegetation, or carton and can be quite elaborate (Andrews 1929; Levieux 1967; Duviard 1969; Duviard & Segeren 1974). Some ant species even build covered runways between the nest and the homopteran shelters (Way 1963). Several studies have shown that wind, rain, or hail may dislodge exposed homopterans, but ants may harvest honeydew from sheltered colonies, even during inclement weather. Some homopterans, especially coccids, inhabit domatia and other nest cavities side by side with ants. Honeydew is often the principal food of species that inhabit domatia (Carroll & Janzen 1973). Perhaps the most protected colonies are the aphid and coccid species, which feed on roots penetrating subterranean nest chambers (Weber 1944). In this closed environment the homopterans are closely guarded, carried to the most succulent roots, and may even have their eggs tended along with those of their ant hosts (Andrews 1929; Landis 1967). On the other hand, Pontin (1978) presented evidence that when root aphids reach high densities, ants such as *Lasius flavus* kill both early instars and adults for food. However, the amount of predation is regulated so as not to interrupt the flow of honeydew to the colony. In one spectacular case studied by Flanders (1957) in central Colombia, the colony size of a root-feeding coccid, possibly of the genus *Eumyrmoccus,* is carefully regulated by the ant *Acropyga* (*Rhizomyrma*) *fuhrmanni,* which determines which individuals have access to the roots. The coccid and ant are obligate mutualists and each queen leaving the nest on her mating flight carries a coccid in her mandibles.

Ants transport homoptera into their shelters or their nests where eggs and juveniles are cared for. They also facilitate feeding by placing them in areas that provide greater access to the phloem of the host (Banks 1962; Way 1963). Some pseudococcids are able to cling to attendant ants and ride to safety when threatened (Reyne 1954). The frequency of this kind of transport is unknown.

The degree of specialization in the relationships between ants and homopterans varies greatly. Jones (1929) presented a mass of data on the relationships between ants and aphids in Colorado. He showed that the average number of aphid species tended per ant subspecies was 7.38, the range being 1–42. The average number of ant attendant species per aphid species was 4.5 with a range of 1–32. Similar arrays or assemblages tending other homopterans have been reported from Scotland (Muir 1959), Iraq (Stary 1969), Ohio (Burns 1964), northern Minnesota (Edinger 1983), and West Africa (Strickland 1947). There is a tendency for homopterans to be tended by only one ant species on any given individual plant, but this is not always so. Furthermore, different ant species have different effects on the homopteran population dynamics, for example, in the degree of protection from predation (Addicott 1979). Finally, many homopterans are able to establish colonies on a variety of plant species (Jones 1929; Muir 1959), although it has been shown that among the Aphidinae of temperate Europe 91% are confined to a single host genus (Eastop 1973).

The services the ants provide – protection, sanitation, and transportation – decrease the development time of individuals and increase the growth rate of the colony, survivorship, and fecundity (Kennedy & Stroyan 1959; El-Ziady 1960; Banks 1962; Way 1963; Banks & Macauley 1967; Bristow 1982). These profound contributions to homopteran fitness, reciprocated by the delivery of honeydew to the ant attendants, form the basis of a vast number of mutualisms, from facultative to obligate, all over the world.

While the homopterans and ants live it up, what is happening to the plants on which they depend? Suckers of plant sap affect host plants in a great variety of ways, disrupting both developmental and metabolic pathways (Osborne 1972). Some either recycle imbibed plant hormones or synthesize them from endogenous metabolites, the net effect being deformation of stems or leaves (Miles & Lloyd 1967). Although abnormalities such as these alter the shape of organs or the form of the plant, the simple extraction of photosynthate from the phloem takes its toll in reducing the size of leaves or the amount of wood formation (Dixon 1971). The former may in turn slow down photosynthesis (Kloft & Ehrhardt 1959) and trigger other subtle but damaging effects such as the loss of abnormal amounts of nitrogenous compounds during seasonal leaf fall (Dixon 1971). Homopterans are frequently the vectors of disease, especially those caused by viruses, which can debilitate or kill the host plants (Weber 1944; Orlob 1963; Maramorosch 1963). The often catastrophic reduction or cessation of flowering and seed set due to homopteran infestations is well known to gardeners, horticulturists, foresters, and farmers.

The damage done to plants by homopterans is very familiar. Anyone who has grown houseplants or tended a vegetable garden speaks only evil of the homopteran. Those who grow plants commercially are even more vituperative and go to great lengths and much expense to eliminate the enemy. With all this folk knowledge and commercial data it seems inconceivable to suggest that homopterans benefit plants in any way. And yet we know that the mere presence of ants can reduce herbivory. Is it possible that so long as homopteran colonies do not grow too large, the protective behavior of the ants may reduce other forms of herbivory and so benefit the plant?

The crux of the matter may be that the overwhelming majority of data on homopteran damage to plants is derived from very artificial situations, most often the monocultures of commercial greenhouses, farmers' fields, or plantations. Population explosions and pest outbreaks are to be expected. The standard response has been to apply pesticides, most of which eliminate any natural predators or parasites and obviate any hope of understanding the natural regulation of homopteran numbers. What really happens in virgin, natural habitats with a full complement of plant, ant, homopteran, herbivore, predator, and parasite species remains virtually unknown.

There are a few tantalizing data. Laine and Niemela (1980) noted that ants tended homopterans on birch (*Betula*) trees in Finland but rapidly turned to active predation during an outbreak of the caterpillar *Oporinia autumnata.* They concluded that the presence of the ants on the trees, in part a result of the homopteran colonies, greatly reduced damage by the leaf-eating caterpillar. Skinner and Whittaker (1981) showed that *Formica rufa* tended aphids on sycamores in England and noted that the trees also supported species of untended aphids and various defoliators, especially lepidopteran larvae. They were unable to assess the balance, if any, between losses due to tended aphids and gains through the removal of other insect herbivores. However, trees with ants lost approximately 1% of the leaf area whereas trees without ants lost as much as 8% of their leaf area. In a very different setting Nickerson et al. (1977) compared the number of eggs of the pestiferous soybean looper (*Pseudoplusia includens*) discovered by ants on soybean plants with and without the membracid *Spissistilus festinus.* Two ant species were tending the membracid, *Solenopsis geminata* and *Conomyrma* (*Dorymyrmex*?) *insana,* both of which were known to eat the looper. The amount of predation on the eggs varied according to time of day, position of plants in the field especially in relation to the principal ant nests, and the position of the eggs on the plants themselves. In one area, dominated by *Solenopsis,* a significantly greater percentage of eggs was taken from plants with membracids. In *Conomyrma-*

Table 6. *Plant characteristics of twenty-six marked stems bearing* Formica *ants throughout the season and their nearest unattended neighbors* ($\bar{x} \pm$ SE)

Plant characteristic	Marked stem with ants	Neighbor without ants	*P*
Initial height at June 30 or July 14 (cm)	91.0 ± 2.8	69.8 ± 2.7	< 0.001*
Seasonal height increment (cm)	16.5 ± 2.0	7.0 ± 2.2	< 0.001*
Number of seeds/stem	3585 ± 671	536 ± 147	< 0.001**

* *t* test.
** Wilcoxon signed-ranks test, due to unequal variances.

dominated areas large numbers of eggs disappeared from both membracid-bearing plants and controls. In all experiments eggs disappeared most rapidly from the lower parts of plants. The data strongly suggested that the membracids attracted ants and that as a result looper eggs were more likely to be found and destroyed, at least on the individual plants that bore the membracids.

In a pioneering study of ant–membracid association on goldenrod (*Solidago*) in New York State, Messina (1981) studied the effects of the ants on the principal herbivores, the chrysomelid beetles *Trirhabda virgata* and *T. borealis.* The ants, chiefly *F. fusca,* tended the membracid *Publilia concava* and would attack the beetles, adults or larvae, whenever they encountered them. The adult beetles would withdraw their legs and drop to the ground to avoid the ants. In order to examine the possible consequences for the plants, two groups of twenty-six plants were marked; the first bore ants and membracids, the second bore membracids only. The stems with ants grew taller and produced a far larger number of seeds (Table 6). In a random sample of stems from the goldenrod population, plants with ants were again taller and maintained a greater leaf area. Surprisingly, plants with ants had almost as many beetle larvae as those without ants. Messina produced evidence and observations suggesting that although ants did bite and even remove larvae, the principal deterrence was to pester the larvae so frequently that feeding was seriously disrupted. Overall, the intensity of ant attendance affected plant growth, the tallest stems being those with the most assiduous ants. Messina concluded that in some circumstances the ant–homopteran interaction did protect the host plant against herbivory.

Messina cautioned that the dynamics were complicated and that a large number of factors could intervene to render the outcome of any given ant–homopteran association uncertain for the plants. Among the factors is density. Banks and Macaulay (1967) found that the presence of low numbers of ant-tended aphids on field beans had little effect on reproductive output. Another factor is the possibility that ants, in protecting homopterans, also deter the predators and parasites of other plant enemies. This is undoubtedly the case sometimes, but Fritz (1982) showed that occasionally predators such as salticid spiders and nabid bugs can successfully avoid ant guards. A further complication is the presence of extrafloral nectaries on plants with ant-tended homopterans. In a study of *Acacia decurrens,* Buckley (1983) evaluated the interactions between a membracid, ants, the extrafloral nectaries, and seed set. In the absence of ants, the membracids are damaging to the plant, decreasing both its growth and seed set. In the presence of ants that guarded the membracids negative effects on the plant were amplified. As the ants constituted effective plant guards in the absence of membracids, it seemed likely that the homopterans disrupted the ant–extrafloral nectary interaction. Buckley proposed that in this system the honeydew was more attractive and rewarding and that the ants abandoned the nectaries to guard the homopterans. In a short time this led either to an increase in the number of membracids sufficient to damage the plant, or to a reduction in the efficiency of defense as the ants clustered exclusively around the homopterans. Interestingly, this did not invariably occur as under some circumstances the ant apparently did not switch to honeydew until so late in the season that seed set was unaffected.

In the previous section on extrafloral nectaries (Chapter 3) I discussed the possibility of wasp guards associated with these glands. Although there are few data on wasp guards associated with homopteran honeydew, they may well turn out to be common. Krombein (1951) spent thirteen afternoons collecting wasps attracted to the large coccid *Toumeyella liriodendri* on tulip trees in Virginia. Incredibly, he collected ninety-three species including three tiphiids, twelve mutillids, seventeen vespids, nineteen pompilids, and forty-two spechids. Among these wasps were many known predators and parasitoids of insect herbivores. Similarly, Zoebelein (1956a, b) reported parasitoids such as ichneumonid wasps and tachinid flies and predators such as coccinellid beetles and syrphid flies feeding on honeydew. The same study showed an increase in the fecundity of some insect species following meals of honeydew. Other authors (Evans & Eberhard [1970]; Vinson [1976]; Price et al. [1980]) have also noted that honeydew attracts a great variety of insects including predatory beetles and wasps and parasitoids. In common with extrafloral nectaries, honeydew-secreting homopterans provide liquid food for a great variety of insects, many of which damage plants either as adults or larvae. Among them are predators and parasites that may provide indirect plant protection.

Lepidopterans

Ants tend the larvae and pupae of a few families of Lepidoptera in much the same fashion as they do their homopteran wards. Larval adaptation to ant tending varies, but there is often an attractive papillalike organ on the dorsal part of the seventh larval segment. This seems to appease ants and inhibit predaceous behavior. The eighth segment usually bears a pair of glands that secrete "nectar" or honeydew when stimulated by the ants. Stimulation generally involves stroking the larvae with the antennae, and the secretions are eagerly consumed by the attendants. The glands are eversible in many species, being retracted when the ant attendant leaves.

The principal Lepidopteran families involved are the Lycaenidae and the Riodinidae comprising a total of perhaps 5000 ant-tended species worldwide (Hinton 1951). Some recent studies have shown that the relationships between caterpillars, pupae, and ants are complex; however, the majority are known only on the basis of caterpillar morphology and the dynamics of the interactions are largely unknown. The chief benefit to the larvae is almost certainly protection from enemies, and the attentions of the ants can be assiduous, maintaining constant and aggressive guard over their caterpillar "cows" (Hinton 1951; Ross 1966; Malicky 1970; Maschwitz, Wust, & Schurian 1975; Atsatt 1981a; Pierce & Mead 1981). Protection from predators and parasitoids may be sufficiently effective to create what Atsatt (1981b) calls "enemy-free space."

The feeding habits of ant-tended larvae are extremely variable and sometimes complex. Caterpillars of the Lycaenidae genus *Lachnocnema* feed on honeydew produced by cicadellid and membracid homopterans. Other genera actually eat the homopterans, and in several cases such as *Laetilia* (Pyralidae), the caterpillar defies ant guards by regurgitating repellent fluids at them (Eisner et al. 1972). A large number of genera are purely phytophagous but *Maculinea* eats plant material only during the first three instars. Larval development is completed in an ant nest where it feeds on the brood. In an ironic twist to this saga, it is the ants themselves that carry the caterpillar into their own nurseries (Hinton 1951).

As with homopterans, ants may build shelters for caterpillars, and the two kinds of honeydew producers are often found within the same shelter (Ross 1966). Adult ant-tended lepidopterans complicate the insect–plant dynamics by feeding on floral nectar, extrafloral nectar, homopteran honeydew, or lepidopteran honeydew, according to availability (Atsatt 1981b).

Again the question arises as to whether the plants benefit from the presence of lepidopterans and ants, and again the data are tantalizing and suggestive. Ross (1966) studied the riodinid larva *Anatole rossi* and showed that ant protection against predators, especially other ants, was so effective that the host plants, *Croton repens,* were occasionally defoliated. Unfortunately the habitat in which the study was carried out was disturbed by frequent man-made fires set to encourage the growth of forage plants for horses and mules.

Another riodinid larva was studied by Horvitz and Schemske (1984) in a less disturbed habitat where species relationships were not subjected to drastic and frequent perturbations. In this case the caterpillar of *Eurybia elvina* ate the flowers of the herbaceous species *Calathea ovandensis* (Marantaceae) and was tended by ants. The ants also visited extrafloral nectaries on the inflorescence. In a series of carefully designed exclusion experiments Horvitz and Schemske found that the plants with no larvae or ants yielded a mean seed set of 17.5 ± 1.2 per inflorescence. By contrast, plants without larvae but with ants yielded a mean of 20.8 ± 1.3 seeds per inflorescence. Although this difference was not statistically significant, it suggested that the ants provided some basic protection against seed predators. The exciting question as to what happened to seed set when the larvae and ants were present was clearly answered. Plants with larvae and ants showed a mean seed set of 13.6 ± 1.1 per inflorescence, which was significantly lower than when the larvae were absent. However, plants with larvae but without ants exhibited the lowest seed set of all, 6.2 ± 0.7 per inflorescence, which was in turn significantly lower than any other. Put another way, *Eurybia* larvae reduced seed set by 33% in the presence of ants, but reduced seed set by 66% in the absence of ants. Clearly the plants set more seed in the presence of ants. The attacks by *Eurybia* may have been ameliorated by the ants in spite of the fact that they tended the larvae. This is because the relationship between the ants and the larvae was somewhat ambivalent. For example, ants would sometimes prevent larvae from settling on an inflorescence, and on other occasions the very attentions of the ants prompted more frequent movements of larvae, apparently reducing their feeding rates. Both of these factors explained in part the higher seed set when both the ants and the predators were present. Other insects such as dipteran larvae also fed on the plants. Ants tending larvae may have also benefited the plants by harrassing all the other plant enemies. In other words, if attacks by seed predators and herbivores were inevitable, it was advantageous to the plant if one type possessed ant attendants that would reduce the deleterious effects of the others. This study provided very important insights into the possible mechanisms for protection of plants by means of ant-tending behavior.

Summary and conclusions

The great preponderance of homopteran literature tells us that they are generally extremely destructive to plants. As a starting point in questioning this general assumption I would like to quote Janzen (1979):

> In at least two complex and well-developed ant-plant mutualisms, African *Barteria* trees (Janzen 1972) and neotropical *Cecropia* trees (Janzen 1973a), the ants maintain a standing crop of scale insects or other homopterans inside the hollow stems. These animals feed on the plant and provide a major food source for the ants with their bodies or their honeydew exudates. The ants are obligate occupiers of the trees and protect the trees from herbivores and vines. The homopterans are zoological devices used by the plants to maintain an ant colony, the ants being directly analogous to the chemical defenses maintained (and paid for) by more ordinary plants. I would not expect there to be selection for traits that reduce the "damage" done by the Homoptera to the level that would debilitate the ant colony and its protection of the tree.

A number of arguments suggest that Janzen was correct.

1. Almost all homopteran studies have been carried out in man-made monocultures or low-diversity systems, that is, greenhouses, crop fields, orchards, and plantations. In such situations where plant, predator, and parasite diversity is low, invaders with appropriate biological attributes (i.e., weeds and pests) become established. In fact, a few field biologists have concluded that homopterans are not particularly common under natural conditions and do not reach the high densities found in man-made situations (Holttum 1954a, b; Way 1968; Eastop 1973; personal observations). If this proves to be the case, some of the costs of homopteran–ant plant defense may be eliminated. Low to moderate densities of homopterans remove concomitantly modest quantities of plant biomass,

the plants appear to be able to withstand some losses of foliage or metabolites before reductions in fitness are detectable (Harris 1973; Mattson & Addy 1975), and in some cases light herbivory may stimulate productivity (McNaughton 1979). If ant-tended homopterans do not reach high densities in natural habitats, then the costs of ant defense may be balanced by the benefits.

2. A few studies have shown that ants may eat the homopterans they tend, whether larvae or adults (Way 1963; Janzen 1972; Carroll & Janzen 1973; Hinton 1977). This behavior may be more frequent when the homopterans reach high densities and is a form of density-dependent regulation (Pontin 1978). Dependence on predation of tended homopterans as a source of protein may be the rule for many ant colonies (Janzen 1972).

3. Southwood (1972) pointed out that Homoptera are highly evolved parasites of plants. In reviewing both the models and experimental work on the evolution of host–parasite interactions, Levin (1983) showed that it usually proceeds toward reduced virulence of the parasite, which leads to minimal harm to the host. In view of this, similar trends are to be expected among homopterans, most of which would mean reduced costs to the plant both in terms of biomass lost, and the burden of maintaining ants for defense. Way and Banks (1967), Way (1968), and van Emden and Way (1972) have demonstrated intraspecific mechanisms for the regulation of numbers of homopterans. Regulation involves slowed reproduction as density increases, increased frequency of emigration, reduction in the size and fecundity of individuals, and sometimes a net decrease in population size, although this is probably associated with the imminent collapse of the host. Mechanisms such as these have probably evolved for the efficient utilization of host plants, but their effects might well lower the impact of homopterans on host plants. If intraspecific mechanisms such as these function in tandem with interspecific mechanisms such as the culling of populations by ants, then homopteran densities may remain low enough for the costs to the plants to be balanced by the benefits of retaining ant guards. If the ant guards remove a significant proportion of the nonhomopteran herbivores, then the benefits may, in fact, greatly outweigh the costs.

4. A few studies such as those of Nickerson et al. (1977), Messina (1981), and Horvitz and Schemske (1984) have demonstrated protection by homopteran/lepidopteran-ant defense systems. It appears that a number of criteria may have to be fulfilled before protection arises from the interaction: (a) The ant-tended partner should not be the primary herbivore; (b) the ants do not allow excessive feeding rates or high densities in

the homopteran/lepidopteran populations; and (c) the ants are very effective in removing nonhomopteran herbivores and seed predators. Variation in these conditions results in variation in the degree of protection. The same studies show that ant species vary with respect to ability to attack and remove nontended herbivores and that there is a strong spatial effect, since proximity to the nest of an effective ant-guard species improves the level of protection. Messina (1981) made the point that the protection of the plant is important to the homopterans or lepidopterans simply because the removal of other plant enemies by ants maintains a relatively healthy host for themselves. Consequently, once the complex conditions of the ant–homopteran–plant interaction are met, all three parties can benefit. Although there is no doubt that the presence of homopterans and lepidopterans have a negative impact on the plant, the presence of ants can have two enormous advantages for the plant: They can regulate the activities and numbers of their charges to some extent, and they can attack and remove the plant enemies that do not secrete honeydew.

5

Myrmecotrophy

When van der Pijl (1955) reviewed the relationships between ants and plants he perpetuated a number of terms, most of which were first proposed by Warburg (1892). These included *myrmecophily* for ant pollination, *myrmecochory* for ant dispersal of seeds, and *myrmecotrophy* for the feeding of ants by plants, principally by means of extrafloral nectaries. Because myrmecotrophy involves much more than the feeding of ants and has largely been replaced by the concept of ant protection of plants, I intend to put the term to another use. If the subject–object relationship of these terms remains consistent, then myrmecophily denotes the relationship in which ants benefit plants by acting as pollen vectors, myrmecochory describes the benefit to the plant conferred by ants dispersing seeds, and myrmecotrophy implies a relationship in which ants benefit plants by feeding them. As far as I can tell most biologists were unaware of this possibility at the time of van der Pijl's review. Since then, however, the existence of ant-fed plants has been documented.

The nutritional benefit to the plants has been clarified for two Rubiaceous genera from Southeast Asia and northern Queensland that house ants in large tubers derived from the embryonic hypocotyl (see Table 7). Janzen (1974b) observed that the ant *Iridomyrmex myrmecodiae* abandons the remains of prey in some of the cavities that ramify the tubers of *Hydnophytum formicarium* and *Myrmecodia tuberosa* (Figure 8). These cavities are lined with absorptive tissues. By contrast, the cavities in which the ants keep their juvenile stages are lined with tough, nonabsorptive suberized cells. Janzen suggested that the chambers served two entirely different functions, the nonabsorptive ones being living quarters while the absorptive chambers are refuse dumps from which the plants absorb the products of decomposition. The latter idea was pursued by Huxley (1978) and Rickson (1979), who performed radioactive tracer experiments to determine the fate of ant refuse in the absorptive chambers. It was immediately apparent that the tuber can absorb (^{32}P) phosphate, (^{35}S) sulphate, and (^{35}S) methionine from materials defecated by the ants, and various (^{14}C) breakdown products of decaying *Drosophila* larvae. These

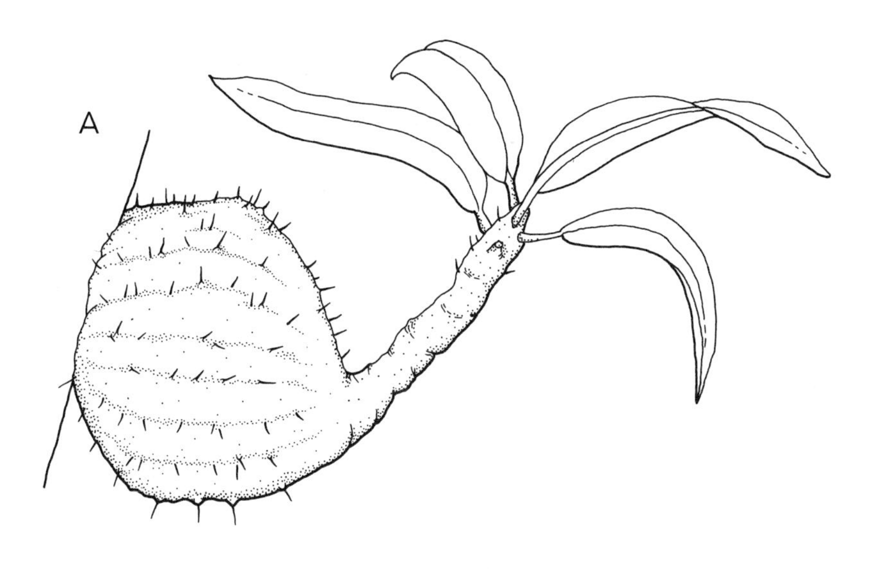

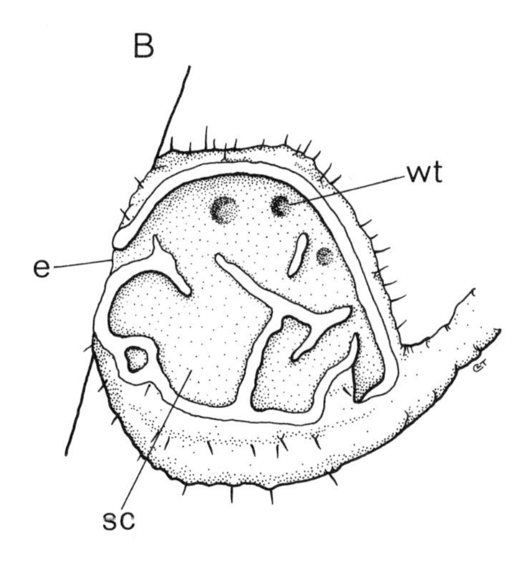

Figure 8. (A) *Myrmecodia tuberosa* Jack, an entire young plant. (B) Whole young tuber cut away to show interior; e, entrance hole; sc, smooth chamber; wt, warted tunnel. (Redrawn from Huxley 1980.)

data demonstrated that ant wastes were incorporated into the tissues of the host plants and, therefore, that the ants fed the plants.

The ecological significance of this was first suggested by Janzen (1974b) and elaborated by Thompson (1981). The proven myrmecotrophic plants

Table 7. *Some plant structures generally associated with ant nests*

Plant structure	Example	Family
Leaf pouch or bladder	*Scaphopetalum***	Sterculiaceae
	*Duroia***	Rubiaceae
	*Remijia***	Rubiaceae
	Ossaea	Melastomaceae
	*Maieta***	Melastomaceae
Swollen petiole	*Nepenthes*	Nepenthaceae
Leaf sheath	*Korthalsia*	Palmae
Pitcher leaf	*Dischidia*	Asclepiodaceae
Swollen stipule	*Uragoga*** (= *Cephaelis*)	Rubiaceae
Inflated leaf bases	*Tillandsia*	Bromeliaceae
	Wittmackia	Bromeliaceae
	Aechmea	Bromeliaceae
Hollow stems*	*Randia***	Rubiaceae
	*Uncaria***	Rubiaceae
	*Sarocephalus***	Rubiaceae
	*Cordia***	Boraginaceae
	*Duroia***	Rubiaceae
	*Patima***	Rubiaceae
	*Clerodendon***	Verbenaceae
Hollow pseudobulb	*Schomburgkia**	Orchidaceae
	*Diacrium**	Orchidaceae
	Caulathron	Orchidaceae
	Epidendrum	Orchidaceae
	Gongora	Orchidaceae
	Vanda	Orchidaceae
Hollow root	*Marckea*	Solanaceae
Rhizome	*Polypodium/ Phymatodes/ Microcorium*	Polypodiaceae
	Lecanopteris	Polypodiaceae
	Solanopteris	Polypodiaceae
Tubers derived from hypocotyl	*Myrmecodia**	Rubiaceae
	*Hydnophytum**	Rubiaceae
Root clusters	*Philodendron*	Araceae
	Anthurium	Araceae
	Nidularium	Bromeliaceae
	Streptocalyx	Bromeliaceae
	Aechmea	Bromeliaceae
	Marckea	Solanaceae

Table 7 *(cont.)*

Plant structure	Example	Family
Root clusters	*Ectozoma*	Piperaceae
	Peperomia	Solanaceae
	Codonanthe	Gesneriaceae
	Phyllocactus/ Epiphyllum	Cactaceae

Note: Not all species in a given genus necessarily bear the structure.
 * Bailey (1924) also lists *Myristica, Kibara, Anthobembix, Pleurothyrium, Humboldtia, Schotia, Platymiscium, Chiscocheton, Endospermum,* and *Gertrudia.*
** Domatia.

are all tropical epiphytes in open forests and savannas growing on nutrient-deficient soils. It is likely that the most limiting nutrient is nitrogen and that its acquisition from the ant waste is the major benefit the plants derive from the ants.

The number of myrmecotrophic species is at present unknown but Thompson (1981) listed 201, of which 94 were species of *Hydnophytum* and 42 were species of *Myrmecodia.* Thirteen species of *Lecanopteris* were included. This Malayan fern develops huge rhizomes that envelop the branches they rest on and are invariably a substrate for the establishment of a mass of other epiphytes. Holttum (1954a) surmised that the huge rhizomes contained a harvest of minerals and nitrogenous materials.

A low-level nutritional benefit may accrue to any plant that has ants nesting in a cavity with an absorptive lining. Decomposing ant waste such as discarded parts of prey are likely to give off ammonia. Hutchinson, Millington, and Peters (1972) found that some plants can absorb significant quantities of ammonia from the air. Porter, Viets, and Hutchinson (1972) showed that labeled ammonia ($^{15}NH_3$) appears in various free amino acids that may be translocated to different parts of the plant. Thus, some ambient ammonia can be metabolized. Janzen (1974b) was the first to suspect that plants benefited from ant nests in this way and went on to speculate that some types of domatia that house unaggressive ants in flimsy structures such as leaf pouches are primarily absorptive surfaces, taking up nutrients from decomposing wastes and nest materials. This general nutritional benefit requires much more investigation especially at the physiological level. As Porter et al. (1972) pointed out,

Table 8. *Effect of presence or absence of ant* Crematogaster longispina *on growth of* Codonanthe crassifolia *in a two-month period*

	Ants present	Ants absent
Number of shoots elongating	50	35
Number of shoots not elongating	12	23

Note: $\chi^2 = 5.98$; $P = 0.025$.
Source: Kleinfeldt (1978).

there is a large gap between merely absorbing nutrients and being able to metabolize them. At the same time, the absorption of ammonium ions incurs less metabolic cost than many other nitrogen sources, provided they do not build up to inhibitory concentrations (Gutschick 1981). Although ammonium ions may be less mobile than, for example, nitrate in the soil, this obstacle may be minimal when ant wastes are packed against absorptive plant surfaces inside moist tubers, leaf pouches, and other domatia. The possibility of low-level myrmecotrophy in a broad variety of ant-inhabited plants is an intriguing but unexplored problem.

A variety of plant structures encourage nesting by ants. Some can be called domatia with certainty, but others appear to be more fortuitous (Table 7). The services provided by ants nesting on or in these plants (if any) are probably varied, including myrmecotrophy, protection, or dispersal, but comparatively little is known about them.

One category of nest site not usually referred to as a domatium is the *Ameisengarten* or ant garden, first described by Ule (1902, 1905, 1906) and later reviewed by Wheeler (1921). (See Table 7 under "root clusters.") These are epiphytic plants of neotropical rain forests dispersed by wind, birds, or ants so that the juvenile plants appear on tree trunks or branches. The young developing roots probe the bark but are mostly exposed to the air. Certain ant species construct nests around them, bringing soil particles and vegetable debris to the young root cluster, which is soon embedded in an organic matrix and surrounded by carton. The process has been studied by Kleinfeldt (1978), who observed nest building by the ant *Crematogaster longispina* on the roots of the epiphytic vine *Codonanthe crassifolia*. It seems that the nest provides a substrate for the vine and that the roots provide some structural support for the nest. *Codonanthe* can survive without ants, but it grows faster and larger with its roots embedded in active ant nests (Table 8).

Codonanthe is associated with at least four genera of ants in Central and South America: *Anochetus, Azteca, Camponotus,* and *Crematogaster.* Some of these ants place seeds in the walls of the nest, which appear to be a nutritive substrate for seedlings. The plant, in other words, is at least partially ant dispersed. Extrafloral nectaries are present, but they occur primarily on the stems rather than the leaves and appear to stimulate the nest-building response, as opposed to protective behavior. *C. longispina* will swarm aggressively if the nest is seriously disturbed, but it is not clear that protection would extend to anything more than the roots of the plant.

Ridley (1910) described ant–plant associations that seem to be more casual than those of the ant gardens. They are typified by the ferns *Platycerium* and *Thamnopteris* (*Asplenium*) in Southeast Asia and northern Queensland. These plants are frequently very large and may form bowls of leaves eight feet across high on the trunks and limbs of rain-forest trees. Ants bring soil and debris to the roots and several kinds nest in the root–debris matrix, for example, *Odontomachus, Pheidole,* and *Dolichoderus.* Table 7 also lists some epiphytic ferns with hollow or labyrinthine rhizomes or tubers that generally contain ant nests. These include species of *Polypodium* and *Lecanopteris* from Malaysia and Indonesia, and *Solanopteris* from the neotropics (Holttum 1954a, b; Wagner 1972). The activities of the ants often lead to an accumulation of soil and debris at the base of the fern, and in *Lecanopteris* particularly, the accumulation becomes a seedbed for a variety of epiphytes. A wide variety of orchids including species of *Cattleya, Coryanthes, Dendrobium, Oncidium, Vanda,* and *Vanilla* possess root clusters or hollow bulbs often inhabited by ants (Jeffrey, Arditti, & Koopowitz 1970).

The activities of ants nesting in these plants may provide a variety of services. Many of these ferns and orchids bear extrafloral nectaries (Jeffrey et al. 1970; Koptur et al. 1982) but whether they attract ant guards is questionable. Ant protection in these plants is mostly inferred from the presence of extrafloral nectaries alone (Jeffrey et al. 1970; Wagner 1972; Madison 1979). The nectaries may attract ants to the plants for other reasons: Seed dispersal by ants is likely in a variety of ant-garden species, and the sori (spore-bearing organs) of some species of the fern genera *Polypodium* and *Lecanopteris* bear elaiosomes and are ant dispersed (Holttum 1954b; Janzen 1974c).

It is probable that the most common effect ants have on plants associated with nests, whether true domatia are present, is to improve the nutritional status of the plant. A great many of the species in Table 7 are epiphytic, and the young roots in particular may be left exposed without the nest-building activities of ants. The simple activity of coating roots with soil particles and debris, and enclosing them with carton, will have an effect analogous to potting a plant. The vine *Dischidia rafflesiana* (Asclepiadaceae) has the capacity to penetrate accumulations of ant wastes in its potlike leaves with small roots.

In summary, myrmecotrophy can take a number of forms. Some of the plant species that bear true domatia such as leaf pouches may benefit chiefly from nutrient accumulation by ants rather than protection. It is not known whether nutritional benefits are low-level or major inputs. Many plant species are associated with ant nests, involving particularly tubers, rhizomes, or root clusters. The provision of moisture, nutrients, or physical protection by ants for these absorptive structures may be important to plant fitness, but much research needs to be done.

6

The dispersal of seeds and fruits by ants

Many species of ants gather seeds. Harvester ants store them in underground granaries and consume them during the winter or dry season. These ants are granivores and the net interaction is usually predation. Other ants gather seeds and fruits distinguished by the presence of external tissues, collectively called elaiosomes, which attract ants and stimulate them to carry the entire seed or fruit back to the nest (Figure 9). There the elaiosomes are removed and typically fed to the larvae. The seeds are then discarded, both intact and viable, either in an abandoned gallery of the nest, or close to an entrance in a refuse pile together with other organic waste. Because the elaiosomes contain ant attractants, and as the seeds are not harmed, the interaction, known as myrmecochory, has long been assumed to be a mutualism. Until recently, however, the advantages of the interaction, especially for the plants, have remained undocumented. Recent studies have shed light on the problem, and there appear to be five current hypotheses on the selective advantage to plants of dispersal of seeds and fruits by ants.

The predator-avoidance hypothesis

Seed predators are so diverse and abundant that plant species must be under great selective pressure to avoid them. The essence of this hypothesis is that seeds released from the parent plant are quickly taken by ants to their nests, where they find refuge from predators. The ants are rewarded with nutritive elaiosome tissue and then rapidly lose interest in the seed, which either is not a part of their diet, or is unavailable because it is protected by a tough seed coat. In the nest the seeds are beyond the sensory capabilities of many predators, and others that might attempt to dig them out are likely to be attacked by the ants in response to the disturbance of the nest.

Ant-dispersed seeds and fruits that bear elaiosomes are eaten by a variety of small mammals and birds (Martin, Zim, & Nelson 1951; Witherby et al. 1952; Heithaus 1981), and by insects such as hemipterans (O'Dowd,

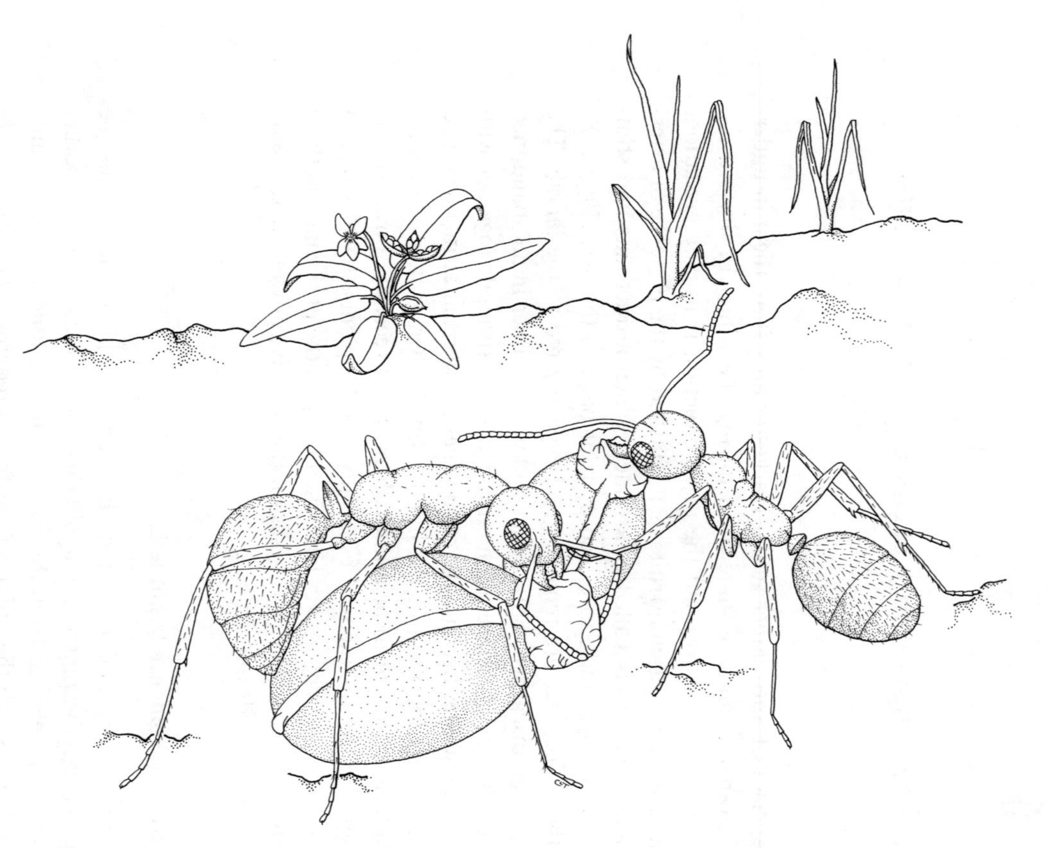

Figure 9. *Formica podzolica* gathering the seeds of *Viola nuttallii*. In each case the ant is manipulating the seed by the elaiosome. A mature plant with an open flower and developing capsule can be seen in the background.

personal communication). Two detailed studies of the fate of ant-dispersed seeds relative to predators have both shown that if the seeds are not discovered by ants within a few hours, they will be eaten (O'Dowd & Hay 1980; Heithaus 1981). The plant species Heithaus studied in the forests of West Virginia experienced very high rates of predation, and ants appeared to be absolutely essential to the survival of seeds and hence the recruitment of seedlings into the population.

The foraging patterns of many ant species are so "thorough" that seeds are generally located within a few hours of release from the parent plant, even if they are scattered individually through the habitat (Culver & Beattie 1978). Some ant-dispersed species merely dump their seeds in a pile and ants recruit workers to the spot until the resource is exhausted (Beattie & Lyons 1975). Others retain the mature seeds for a short period and ants ascend the plants to harvest them directly from the fruit (Gislen 1949; Berg 1954; Davidson & Morton 1981b). Seeds may remain vulnerable to predators if they are returned to waste piles or graveyards on the surface. However, when this happens they may be quickly reinterred (Horvitz & Beattie 1980), or they may be less attractive to predators following the removal of the elaiosomes by ants. A few are discovered by predators and eaten (Turnbull, personal communication). Although ant manipulation may not be foolproof, the evidence is that it can be effective.

The competition-avoidance hypothesis

The competition-avoidance hypothesis is the work of Handel (1976, 1978), who studied three species of *Carex* growing together in the eastern forests of the United States. Two, *C. platyphylla* and *C. plantiginea,* were dispersed by other means. Greenhouse experiments showed that the ant-dispersed species, *C. pedunculata,* did not grow well in the immediate presence of the other two. These data, together with extensive field studies, strongly suggested that interspecific competition for seedling microsites was substantially reduced by ant dispersal. Ants took *C. pedunculata* seeds to microsites (nests) that neither of the other species could exploit. Thus free from competition from its sympatric congeners, it grew vigorously. A small number of other ant-dispersed/non-ant-dispersed congeneric groups of species have been reported (Ulbrich 1939; Uphof 1942), and so this type of selective advantage may occur in genera besides *Carex.*

The fire-avoidance hypothesis

The fire-avoidance hypothesis was primarily developed by Berg, who has led research into ant-dispersed plant species of fire-climax communities. He showed that myrmecochory is common in the dry sclerophyllous shrublands of many parts of Australia, there being as many as 1500 plant species involved (Berg 1975). Since then Milewski and Bond (1982) and Bond and Slingsby (1983) have examined similar, fire-prone vegetation known as *fynbos* in the Cape Province of South Africa and estimate that as many as 1300 plant species may be ant dispersed there. The hypothesis states that seeds taken into nests remain safe from incineration during forest fires. In this context myrmecochory becomes one of a variety of interesting mechanisms for the regeneration of vegetation following fire. However, many ant-dispersed species in fire-prone environments may require fire to stimulate germination. Shea, McCormick, and Portlock (1979) showed that some ant-dispersed species from Western Australia actually require the high temperatures of fires to stimulate germination. These findings suggested that if seeds are to benefit from ant manipulation, the average depth of burial would have to be such that the seeds received sufficient heat to germinate, without being killed. Majer (1982) has since produced evidence that this may be what happens. Working in southern Western Australia with species of *Melophorus* and *Rhytidoponera,* he analyzed the distribution of burial depths relative to seedling emergence. He found that for ant-dispersed plant species at least, germination and seedling emergence was far more frequent on ant nests than anywhere else. The seeds of two species germinated more frequently when nests were heated with an infrared heater in an attempt to mimic the effects of fire. However, five species showed reduced germination under this treatment. Field experiments demonstrated the burial depths that yielded the greatest amounts of germination and seedling emergence. In general, whereas these varied both with the ant species and the plant species, most seed was buried at depths suitable for germination and postfire regeneration. Similar findings were reported by Bond and Slingsby (1983) for the Cape Province, where seeds were concentrated in nest galleries 4–7 cm below the surface. As ant-dispersed species appear to be very abundant in some fire-dependent communities, this selective advantage may be crucial in some parts of the world.

The dispersal-for-distance hypothesis

If juvenile mortality is inversely related to dispersal distance, then the simple transportation of seeds by ants across significant distances may be the primary selective advantage of the seed–ant interaction. Westoby and Rice (1981) developed a model to explain why some Australian ant-dispersed species are diplochorous (i.e., they have a two-stage dispersal system). In most cases seed is hurled explosively from the parent plant and then discovered by the ants. Westoby and Rice showed that whereas many such seeds would be carried back toward the parent plant, about 50% would actually be carried even farther away. If seedling survivorship was limited to a region beyond the sphere of influence of adult plants, selection for ant manipulation might be important. This advantage may be especially critical for ant-dispersed species that form large clones. Luond and Luond (1981) showed that ants generally take seeds beyond the boundaries of parent clones, thus eliminating seedling competition with established adults and increasing the probability of seedling survival. Pudlo, Beattie, and Culver (1980) also showed that in disturbed habitats where seed-dispersing ant species were absent, the clones persisted but seedling establishment was virtually eliminated. Seeds not taken beyond the clones simply succumbed to competition with the vigorous parental ramets.

The nutrient hypothesis

Ant nests in the ground are generally distinct from the surrounding soils both chemically and physically. For example, nests may differ from adjacent soils according to temperature, porosity, moisture levels, pH, organic content, a wide array of minerals, and the diversity and abundance of microorganisms (Talbot 1953; Scherba 1962; Petal, Jakubczyk, & Wojcik 1967; Wilson 1971; Gentry & Stiritz 1972; Malozemova & Koruma 1973; Haines 1975; Wells et al. 1976; King 1977a, b, c; Petal 1978). The essence of this hypothesis is that ant nests are especially rich in some or all essential plant nutrients. Thus, seeds taken into nests arrive at better microsites for germination and establishment than if they had been scattered at random across the surrounding soils. Put another way, ants create fertilized pockets of substrate that the plants locate by making their seeds or fruits attractive to ants. The elevated levels of nutrients are derived from the activity of the colony, particularly the accumulation of waste materials and the formation of graveyards.

A recent study of the nest chemistry of a seed-dispersing ant species supports this hypothesis. Beattie and Culver (1983) analyzed the nests of *Myrmica discontinua* near the Rocky Mountain Biological Laboratory at 9500 feet in the mountains of southwestern Colorado. This ant species is common in the montane meadows of the region, which are rich in herbaceous perennial species, including at least four that are ant-dispersed: *Viola nuttallii, Delphinium nelsoni, Claytonia lanceolata,* and *Mertensia fusiformis* (Turnbull, Beattie, & Hanzawa 1983).

The nests of *M. discontinua* are entirely subterranean, often beneath a log or rock, with entrance holes frequently concealed among the stems of plants. In addition, a network of surface tunnels radiates from the nest so that workers may travel several meters before actually emerging into the open (Christine Turnbull, personal communication). Foragers are difficult to follow, being cryptic and prone to utilize the surface tunnels.

Samples of soil were taken from the center of *Myrmica* nests and compared to samples taken at random from surrounding soils. Fifteen soil variables were analyzed and these, together with the results of the analyses, are shown in Table 9.

The nests differ chemically from the surrounding soils in showing elevated levels of the macronutrients phosphorus and nitrate (ammonium). The nests also showed more cadmium, less nickel, and more organic matter than surrounding soils. These differences were further investigated by discriminant function analysis in which the existence of the two groups was assumed (i.e., *Myrmica* nests and controls), and the samples classified according to these groups. The first discriminant function was very powerful, accounting for 91.7% of total variance, and the first three together separated nests and controls with only two misclassifications (Beattie & Culver 1983).

Assessment of the hypotheses

Avoidance of predation may be crucial in certain habitats but of less importance elsewhere. Rodents are important seed predators of myrmecochorous plant species in deciduous forests of the eastern United States. However, experiments with elaiosome-bearing seeds in secondary-growth rain forest in southern Mexico, and in semiarid scrubland in Australia, suggest that rodents in these habitats are not a major danger to seeds (Horvitz, personal communication; Davidson & Morton 1981b). A variety of other animals, especially birds, hemipterans, and other ants, may prove to be important predators in some habitats, but there are insuffi-

Table 9. *Means and standard errors of fifteen soil variables in the nests of* M. discontinua

	M. discontinua nests		
Soil variable	$\bar{X}$	SE	More/less?
Phosphorus (ppm)	2.93	0.45	More (0.002)
Potassium (ppm)	260	28	n.s.
Zinc (ppm)	3.31	0.26	n.s.
Iron (ppm)	118	10	n.s.
Manganese (ppm)	16.9	1.60	n.s.
Cadmium (ppm)	0.26	0.01	More (0.05)
Lead (ppm)	2.79	0.11	n.s.
Nickel (ppm)	0.45	0.04	Less (0.001)
NH_4-N (ppm)	5.67	0.57	More (0.05)
Organic matter (%)	7.73	0.42	More (0.05)
Copper (ppm)	2.18	0.14	n.s.
pH	6.47	0.05	n.s.
Conductivity	0.33	0.12	n.s.
Sodium absorption rate	0.17	0.06	n.s.
NO_3-N (ppm)	1.90	0.30	n.s.

Note: The third column indicates the level relative to the controls from surrounding soils, and the numbers in parentheses give the significance of the *t*-test for differences between nests and controls.

cient data to claim that predator avoidance is a general benefit derived from ant dispersal.

The rationale of the competition-avoidance hypothesis is limited to congeneric species with and without elaiosomes that grow in the same habitat. If other ant-dispersed species are also present, competition for nest microsites may be as formidable and unavoidable as for any other microsites.

The contribution of ant dispersal to fire avoidance remains somewhat of a mystery. Majer's (1982) experiments showed that seedling emergence is possible away from ant nests and that some levels of heat destroy seeds whereas others promote germination. Relocation of seeds to nests may increase the probability of surviving fire in some plant species, but the situation could be more complex. For example, O'Dowd (personal communication) suggests that the seeds of some species may be able to survive low-intensity fires outside ant nests, but require burial by ants to survive high-intensity fires. Ant dispersal thus becomes a form of "bet hedging" in anticipation of the occasional devastatingly hot fire. The fire-avoidance hypothesis is limited in scope as it applies only to ant-dispersed species in fire-dependent vegetation.

The dispersal-for-distance hypothesis simply lacks supporting data and may be confounded by arguments that can be plausibly advanced for almost any dispersal distance. For example, short dispersal distances may be optimal as the presence of a seed-producing parent is reasonable evidence that the immediate environment may be suitable for establishment. On the other hand, it has been argued that concentrations of seeds close to parent plants are highly vulnerable to predators and that intermediate or even long-distance dispersal avoids this problem, and also removes the seedling from competition with the parents. Longer distances may also facilitate colonization of unutilized habitat. Dispersal distances of ant-borne seeds vary from a few millimeters to over 100 meters. Frequently, the seeds of a single plant population are carried by a variety of ant species and the distribution of dispersal distances can be leptokurtic or platykurtic (Beattie & Culver 1979; Wyatt & Stoneburner 1981). Much work has to be done on this aspect of dispersal in general. In particular, there is the question of whether selection favors the entire distribution of dispersal distances, which implies selection for arrays of ant dispersers, or whether selection favors one type of disperser (ant species) over another as this may optimize the frequency of an advantageous subset of dispersal distances (e.g., large ants taking seeds long distances).

The nutrient-enrichment hypothesis is the most predictive of all. The elevated levels of macronutrients in the nests are in broad agreement with the findings of other studies. For example, phosphorus and potassium are more abundant in the nest of *Formica* species such as *F. cunicularia* (Czerwinski, Jakubczyk, & Petal 1971), *F. lugubris, F. polyctena,* and *F. pratensis* (Malozemova & Koruma 1973), and *F. rufibarbis* (Petal 1978). These elevated levels are greatest in the center of the mound (Malozemova & Koruma 1973) and may persist even after the nest has been abandoned for several months (Petal 1978). The nest chemistry of other species of *Myrmica* is similar to that of *M. discontinua* with elevated levels of phosphorus, potassium, and nitrate (Czerwinski et al. 1969, 1971; Petal 1978). Other seed-dispersing ant genera follow suit; for example, elevated levels of potassium, phosphorus, and nitrogen are found in *Lasius* nests (Czerwinski et al. 1971), and the mounds of some *Rhytidoponera* species in Australia are especially rich in phosphorus and nitrogen (Davidson & Morton 1981b). In summary, the nests of seed-dispersing ant species do exhibit higher levels of some essential plant macronutrients than surrounding soils.

The picture for micronutrients and other minerals is far more variable and confused. The nests of *F. canadensis,* for example, show diminished levels of some micronutrients compared to adjacent soils (Culver & Beattie 1983). However, there was no evidence that they were deficient relative to normal plant growth (Soltanpour, Ludwick, & Reuss 1979).

Three heavy metals - cadmium, lead, and nickel - were also at low levels in the nest of *Formica.* By contrast, cadmium was at an elevated level in *Myrmica* nests, a puzzling situation compounded by a similar finding for the nests of the seed-dispersing ant species *Pachycondyla harpax* in secondary rain forest in southern Mexico (Carol Horvitz, personal communication). It is unlikely that the levels of cadmium found in ant nests are harmful to plants, particularly as the synergistic action of potassium appears to increase plant tolerance to cadmium (Simon 1977). There are also hints in the plant physiology literature that cadmium may not be as harmful to plants as previously supposed and may even be beneficial to some (Leavitt, Dueser, & Goodell 1979; Brown & Martin 1981). This discussion illustrates the complex chemistry of ant nests; however, the data support the contention that they are richer than surrounding soils in some essential plant nutrients.

Unfortunately, no experiments have been performed to specifically test the nutrient-enrichment hypothesis. The nearest attempt simply asked the question: Do seedlings that emerge from nests grow better than those that emerge from other microsites? When violet seeds are taken by ants, three things happen: (1) They are gathered individually from the habitat and aggregated into small groups in the nest; (2) the elaiosome is removed and the seed coat gnawed or scarified; and (3) a proportion of the seeds germinate and emerge from the nest (Culver & Beattie 1978). Working with two species of ant-dispersed violets on the chalk downland of southern England, Culver and Beattie (1980) performed a series of planting experiments to analyze the relative importance of each of these effects. The protocol of the experiments can be summarized as follows:

Experiment	*Ant effect(s) tested*
1. Seeds planted singly, by hand, at random	None
2. Seeds planted in groups of ten, by hand, at random	Grouping
3. Seeds scarified and planted by hand in groups of ten, at random	Grouping and scarification
4. Groups of ten seeds relocated to ant nest by undisturbed ant foragers	All effects of ant nest

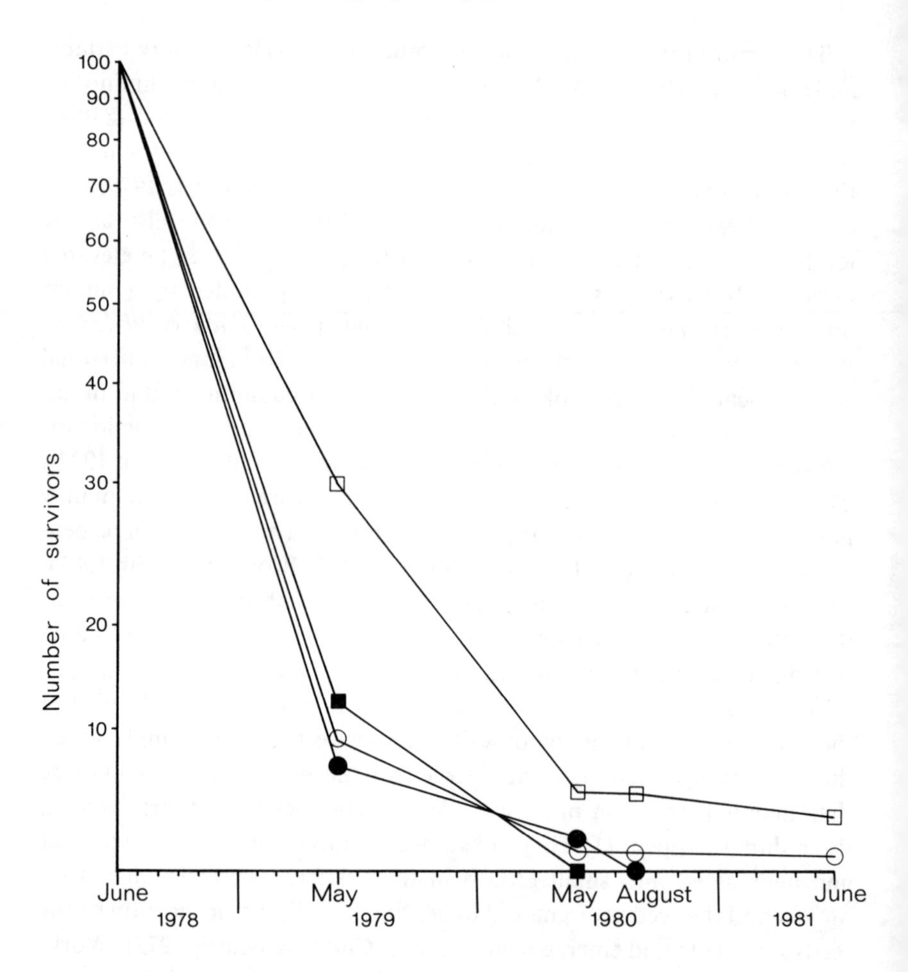

Figure 10. Survivorship of *Viola* seedlings emerging from planting experiments. Open squares, seedlings emerging from ant nests; closed squares, seedlings in the random treatment; open circles, clumped and scarified treatment; and closed circles, clumped treatment. The experiments were carried out over 36 months in southern England. (From Beattie 1983.)

Each nest and planting site was monitored for thirty-six months. Seedlings that emerged from nests were the most numerous and were larger than those that emerged from random microsites (Culver & Beattie 1980). At the end of three years, survivorship was highest for the nest emergents (0.04), most other seedlings having expired by the second year (Figure 10).

The single survivor from the random treatments was from the "scarified and grouped" category, which most closely imitated the other important effects of ant behavior. Germination and emergence had occurred at microsites other than nests but did not contribute any significant survivorship. The numbers of survivors were very low but were of the same order of magnitude as other violet populations (Cook 1979, 1980; Solbrig, Newell, & Kincaid 1980; Solbrig 1981) and various ant-dispersed herbaceous perennials (Tamm 1956; Kawano 1975; Ernst 1980; Kawano, Hiratsuka, & Hayashi 1982).

There is a possibility that the accumulation of seeds in nests leads ultimately to seedling competition. There are several reasons for believing that this does not occur. First, we saw no sign of it in the violet experiments. Second, many of the ant species that disperse seeds frequently relocate their colonies, in some cases every three to four weeks (Colombel 1970; Petal 1978; Smallwood & Culver 1979; Smallwood 1982b). This would militate against large accumulations of seeds in nests. Third, there is some phenological displacement of seed release among guilds of ant-dispersed species (Beattie & Culver 1981; Judy Smallwood, personal communication) so that the buildup of more than one plant species per nest may be minimized. The violet experiments suggest that even if several seeds are planted together (groups of ten were used), the probability of emergence from ant nests is still greater than elsewhere. These experiments strongly suggest that nests are superior microsites for emergence and establishment. However, they do not identify nutrient enrichment as the factor involved, and with the present state of knowledge, this can only be inferred from the chemical data outlined in this chapter.

The dynamics of some chenopodiaceous ant-dispersed species taken by the ant *Rhytidoponera* in eastern Australia have something in common with those of the violet populations, but also differ in some fundamental aspects. Davidson and Morton (1981a, b) found that *Rhytidoponera* mounds contained a hundred times more inorganic nitrogen than surrounding soils. The nests also exhibit elevated levels of phosphorus and significantly less soil compaction than random samples taken from the immediate area. Elevated nutrient levels appear to be the chief advantage of ant dispersal in this situation. Here the possible resemblance to the violet ends as there is clearly intense competition among ant-dispersed species for relocation to nutrient-rich mounds. Individuals of ant-dispersed species on nest mounds are more numerous, grow larger, and live longer than those existing on the surrounding soils.

The nutrient hypothesis explains much of the variation in abundance and diversity of ant-dispersed species on a worldwide basis and has, therefore, a greater predictive power than any of the other hypotheses. Ant-dispersed species seem to be most diverse and abundant on poor soils, especially those deficient in phosphorus, potassium, or nitrogen. Two major types of environment are well known to harbor large numbers of ant-dispersed species: the temperate forests of Europe and eastern North America, and various kinds of sclerophyll shrublands in Australia and South Africa. I shall discuss these in turn.

A large proportion of the nutrients in temperate forests is sequestered in the tree biomass and are consequently of limited availability to the herbaceous plants on the forest floor (Grime 1979). Other nutrients are leached from the soil through precipitation and drainage (Likens et al. 1977). The end result is that some nutrients, especially phosphorus and nitrogen, are in limited supply, at least for the herbaceous layer (Whittaker 1975). A similar phenomenon is well known in various kinds of tropical forests where the situation is often exacerbated by the presence of ancient soils already poor in plant nutrients (Richards 1966; Janos 1980). The availability of nutrients for small plants on the forest floor is both brief and patchy seasonally and spatially (Muller & Bormann 1976; Muller 1978; Vitousek et al. 1979). As the violet experiment showed, germination can occur in many places, but the establishment of seedlings is favored at ant nests. These structures have the dual advantage of being richer in nutrients than the surrounding soil, and being somewhat buffered with regard to the source of supply as this rests with the waste-accumulating activities of ants, rather than direct nutrient competition with the dominant trees.

The relocation of seeds to microsites already provisioned with nutrients is likely to be particularly advantageous to plant species in soils where nutrients are limiting, either because of a lack of parent materials or because of competition. However, this advantage is not confined to plants growing on nutrient-deficient soils, and may accrue to plants with exceptionally high nutrient demands on more fertile soils. Some spring ephemerals of deciduous forests of the eastern United States fall into this category. Leaf and flower production takes place in a few weeks between the end of winter and the closing of the tree canopy. The rapid growth and high photosynthetic rate required for this are manifested, in part, by a very high nutrient demand, especially for nitrogen (Givnish, Terborgh, & Waller in press). This demand is crucial for seedlings and they would be at a selective advantage if emerging predominantly from ant nests.

The shortage of nutrients available to ant-dispersed species in temperate forests is primarily the result of competition from the dominant tree species. The situation in many parts of Australia and South Africa is similar in that nutrients are in short supply. However, the most fundamental cause may be a lack of parent materials on which competition for limiting nutrients is superimposed. Many Australian soils are notoriously old and impoverished, being deficient in phosphorus, potassium, nitrogen, and an array of micronutrients (Leeper 1949; Wild 1958; Beadle 1962; Beard 1976). In a very interesting comparison of the incidence of ant-dispersed species on fertile and infertile soils in southwestern Australia and the southwestern tip of South Africa, Milewski and Bond (1982) concluded that soil type was the most important factor involved. Soils deficient in phosphorus, exchangeable potassium, calcium, magnesium, sodium, and total nitrogen harbored significantly more species of ant-dispersed plants than did more fertile soils. This has also been documented for other parts of Australia (Davidson & Morton 1981b; Drake 1981). Milewski and Bond (1982) demonstrated the contrast with other Mediterranean, semiarid, and arid areas around the world that have more fertile soils. In these areas ant-dispersed plant species are far less abundant. Their work pointed to the conclusion that in many habitats ant dispersal is a response to nutrient deficiency and that the end result is the repeated appearance of a suite of adaptations to ants among plant species in disparate floras around the world.

In Australia ant-dispersed plant species are very common on some soil types and rare on others. In seeking an explanation for this Westoby et al. (1982) made three interesting discoveries: First, ant-dispersed plants tend to be common on soils with very low levels of phosphorus; second, chemical analysis of the elaiosomes of two myrmecochores from phosphorus-deficient soils revealed a very low phosphorus content; third, a similar analysis of the wings of the seeds from five wind-dispersed species growing on the same soils revealed the presence of little phosphorus. These data prompted the suggestion that when phosphorus is limiting, it is advantageous to possess a dispersal mechanism that requires only minimal quantities. As both the elaiosomes and wings contained minimal amounts, phosphorus was viewed as the resource "currency" that controlled the evolution of the dispersal mechanisms, and the two types of dispersal analyzed were regarded as "cheap" since they apparently required little phosphorus. Westoby et al. (1982) argued that limiting factors or resource currencies such as mineral nutrients might explain the distribution of many kinds of dispersal mechanisms, and predict their relative frequencies in different vegetation types on different soils.

It is to be hoped that more data are gathered to test this interesting hypothesis, especially for the phosphorus-deficient Australian habitats where the work began. For several reasons, it seems likely that a more complex picture than the one originally proposed will appear. First, the dispersal mechanisms were not analyzed for the dozen or more other macro- and micronutrients that might be limiting factors, or might contribute to the limiting properties of phosphorus. Second, phosphorus is generally at low concentrations in plant tissues with the frequent exception of fruits and seeds, and the concentrations of this element found in the elaiosomes were not unusually low and even fell in the upper range for some plant tissue analyses (Ovington 1956; Sicama, Bormann, & Likens 1970; Bidwell 1974; Woodwell, Whittaker, & Houghton 1975; Raven, Evert, & Curtis 1981). Third, phosphorus-deficient soils are extremely common worldwide, but many such as serpentine soils do not support the high frequencies of myrmecochorous species such as those found in Australia (Walker 1954; Whittaker 1954).

There is little doubt that the supply of phosphorus and nitrogen is crucial to plant rewards and ant services. However, the reality of a "currency" or limiting factor is best established by the discovery that a given ant service is regulated in quality or quantity by the supply of a given nutrient in the plant. In Chapter 8 it will be seen that rewards such as extrafloral nectar and honeydew vary greatly in composition. What we do not know is whether ants fail to take rewards, or provide services, if a given element is in short supply. We also do not know whether ants can make up for such shortages from other sources. In addition to this, selection for ant services may be sufficiently intense to divert certain nutrients to ant rewards even though they are in short supply. For example, plant tissues and plant sap are generally low in nitrogen content (Slansky & Feeny 1977; McNeill & Southwood 1978; Mattson 1980), and yet nitrogenous substances, especially amino acids, are prominent in all ant rewards (Chapter 8). By contrast, phosphorus appears to be unusual in ant rewards, the single exception being a phospholipid.

In summary, plants offer nitrogen in ant rewards and, in the case of seed dispersal, ants generate nitrogen-rich germination and establishment sites for plants. The degree to which the nitrogen is recycled between the mutualists is unknown. Phosphorus is different in that it rarely appears in ant rewards, and yet in the case of seed dispersal, the ants may concentrate the element at germination sites.

It should be emphasized that ant nests may have other advantages besides nutrient availability. Nest soils may be better aerated, less prone to dry out, or may experience fewer fluctuations in temperature (see preceding references). In some temperate habitats with dense vegetation the successful nests are those that experience moderate levels of insolation, being neither too shaded nor too exposed (Elmes & Wardlaw 1982; Smallwood 1982a). Seedlings derived from seeds taken to such nests may be at a similar advantage.

The distribution and variability of ant dispersal

If nutrient deficiency is indeed central to the evolution of ant dispersal, then concentrations of ant-dispersed species can be anticipated in other plant communities. The first is surely tropical rain forests where two kinds of plants appear to experience nutrient limitation, namely, forest floor herbs and epiphytes. Horvitz and Beattie (1980) and Lu and Mesler (1981) have identified several forest floor species in neotropical rain forests that are ant dispersed. N. B. M. Brantjes (personal communication) believes there are many more in the rain forests of Brazil. As for epiphytes, Ule (1902, 1905, 1906) reported several from the forests of Guyana. These reports were disputed by Wheeler (1921) and Weber (1943) but van der Pijl (1972) and Kleinfeldt (1978) have since shown that Ule was correct. In addition, Wilhelm Barthlott of the University of Berlin has been investigating the minute seeds of *Blossfeldia, Frailea,* and *Setiechinopsis,* three genera of epiphytic cacti. These seeds are generally considered to be wind dispersed. However, under the microscope the organs formerly considered wings have turned out to be moist, and highly attractive to the minute ant *Tapinoma melanocephalum.* This ant gathered seeds avidly in experimental cages, returning them to the nest, chewing off the "wings" (elaiosomes?), and feeding the tissue to larvae. As the seeds were not harmed, the behavior appeared to be routine ant dispersal. Although this ant species may not be present in the habitats of the cacti involved, many others of the appropriate size certainly are. In Chapter 5 I reviewed a variety of other epiphytes, including ferns, probably dispersed by ants.

A second type of habitat where concentrations of ant-dispersed plants might be anticipated is the various grasslands and savannas where soils are deficient in plant nutrients. Bews (1917) wrote of parts of South Africa,

"...preliminary investigations are sufficient to show that ants are important agents for seed dispersal in the veld." Van der Pijl (1955, 1972) anticipated that ant dispersal would be important in some grasslands as far apart as Argentina and Java. To my knowledge this has not yet been confirmed or denied.

Ant-dispersed plant species have been reported from many other kinds of vegetation and include growth forms as varied as small trees, shrubs, herbaceous annuals and perennials, vines, epiphytes, hemiparasites, and parasites (Beattie 1983). Among the approximately seventy plant families containing ant-dispersed species, the following are especially important: Berberidaceae, Boraginaceae, Compositae, Epacridaceae, Euphorbiaceae, Fabaceae, Fumariaceae, Iridaceae, Liliaceae, Mimosaceae, Papaveraceae, Proteaceae, Ranunculaceae, Sterculiaceae, Turneraceae, Violaceae, and Zingiberaceae. Ant dispersal is also found in several species of grasses (Graminae), sedges (Cyperaceae), and cacti (Cactaceae).

Ant-dispersed plants are often a significant part of a local flora. Sernander (1906) showed that in some habitats in Sweden 40% of the herbaceous species were myrmecochorous. At high altitudes in Siberia he gave a figure of 6%, and at high altitudes in the French Alps and in the northern Tyrol the figures were 7% and 13%, respectively. In a variety of forests in the eastern United States, ant-dispersed species comprise 26%–35% of the species present in the herbaceous layer, and up to 76% of emergent herbaceous stems (Beattie & Culver 1981; Handel et al. 1981). The sclerophyll forests and woodlands of part of New South Wales have between 24% and 37% myrmecochorous species (Westoby et al. 1981), and this rises to 50% or more among the dominant shrubs (Berg 1975). Elaiosomes take many forms (see Figure 11). In temperate plants they tend to be soft and fleshy, and usually decay within a week or two if not eaten by ants. By contrast, the elaiosomes of plants in arid areas are tough, resistant, and may persist for weeks or months.

Elaiosomes often appear on fruits or seeds that possess obvious adaptations to additional means of dispersal. For example, the seeds of *Viola* are initially dispersed by ballistic ejection from a capsule before being discovered by ants (Beattie & Lyons 1975), and this is an extremely common combination of dispersal methods (Westoby et al. 1982). In ant-dispersed species of *Reseda* the seeds are first shaken out passively, and in some Compositae the fruit actually bears a pappus or wings at one end and an elaiosome at the other (Nesom 1981). In one species of *Trillium* the red berries are attractive to birds, but if they are not taken the outer wall falls away to reveal seeds with elaiosomes. Although many of these

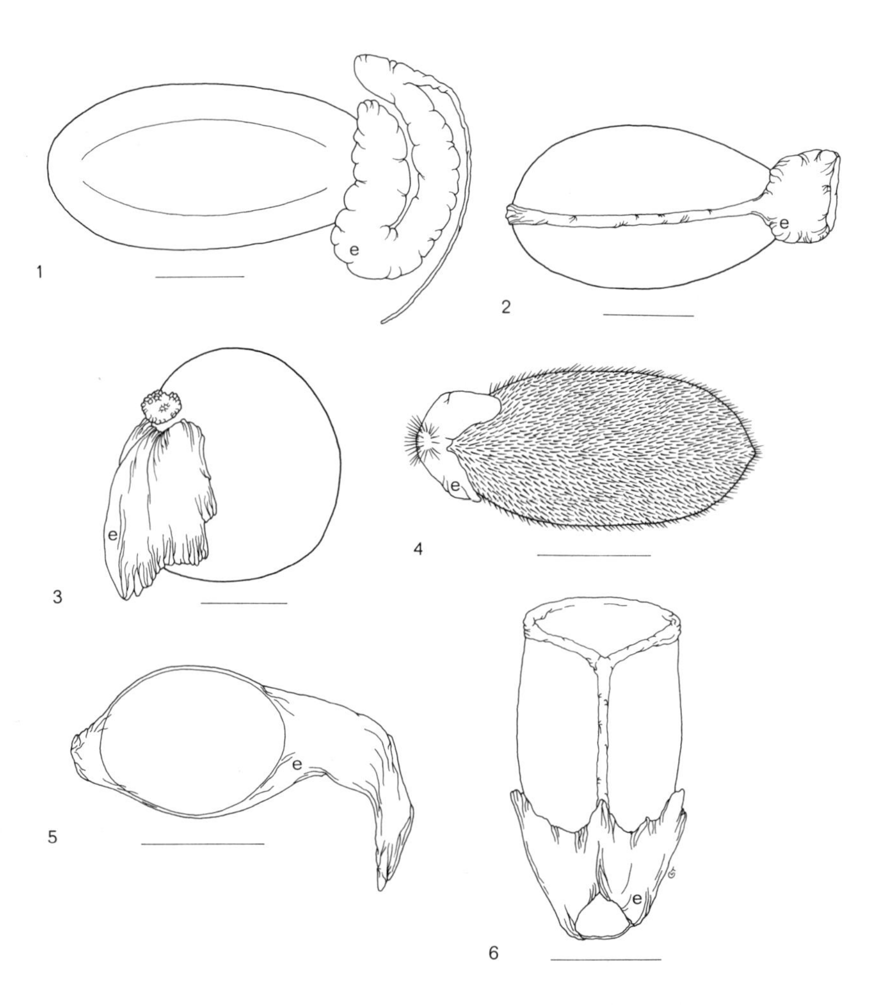

Figure 11. Elaiosomes (e) on the diaspores of the following plant species: (1) *Acacia verticillata,* (2) *Viola nuttallii,* (3) *Corydalis aurea,* (4) *Polygala vulgaris,* (5) *Luzula campestris,* and (6) *Lamium album.* The scale beneath each diaspore represents 1 mm.

two-stage (diplochorous) systems exist, it is very common for ant-dispersed species to simply dump their seeds in piles immediately below the parent plant.

The anatomical and morphological differences between elaiosomes and the larger appendages called "arils" are often subtle and confusing. This is reinforced by the dispersal agents, which are thought to be ants in the case of elaiosomes and birds in the case of arils. Some "arillate" diaspores from forest floor plants in the neotropics have been seen to be taken by birds at some places and ants at others. Davidson (unpublished data), working with plants in arid areas of Australia, believes that some are adapted to both ant and bird dispersal and sees no reason why the plants should not have evolved with both types of dispersal agent. The abundance and variety of two-stage systems suggest that in many cases dispersal is a two-stage process: displacement and inhumation. Mechanisms such as ballistic ejection, the shaking of the plant, wind, and travel in the gut of a bird all contribute to displacement from the parent plant. This does not necessarily involve or require movement across long distances, but rather the creation of some optimal distribution or "shadow" of seeds around the parent plant or population. The second stage, inhumation (from Latin *humare,* to cover with earth), is one in which seeds come to rest in a protective or nutritive microsite, such as an ant nest (Beattie & Culver 1982). This process has been described for a number of vertebrate seed-dispersal agents that defecate intact seeds, which germinate directly from the dung (Ridley 1930; McKey 1975; Applegate et al. 1979; Lieberman et al. 1979; Janzen 1981; Rogers & Applegate 1983). Beattie and Culver (1982) have suggested that inhumation by ants, which are far more abundant and ubiquitous than vertebrate seed dispersers, is a vital second-stage process to many plant species in diverse habitats worldwide. They went on to suggest that other common invertebrate groups such as beetles of the family Scarabidae, earthworms, and some kinds of snails may also inhume seeds. Many kinds of seeds and fruits, including those bearing elaiosomes, do not simply come to rest at random locations in the habitat once they are released from the parent plant. Ants, and possibly other invertebrates, relocate them to sites that are protected and/or fertilized. Because microsites suitable for germination and establishment are often patchy and scarce (Harper 1977; Solbrig et al. 1979), selection for seed relocation to superior microsites is intense.

The ants that disperse seeds

The ant species known to disperse seeds and fruits all belong to one of four subfamilies of the Formicidae: the Formicinae, the Myrmicinae, the Ponerinae, and the Dolichoderinae. In a recent review I identified thirty-two genera as having seed-dispersing species (Beattie 1983), but several more studies have since been done and the list is growing rapidly. The four subfamilies with seed-dispersing species are notable for their variety of nest types, foraging patterns, and food sources. This is in contrast, for example, with the subfamily of Leptanillinae (one of subterranean raiding ants), the subfamily Ecitoninae (the famous legionary ants), or the Pseudomyrmecinae (which dwell in hollow stems). Many of the seed-dispersing genera have large geographical distributions, for example *Pachycondyla, Pheidole, Odontomachus,* and *Crematogaster* are "tropicopolitan"; that is, they occur throughout most of the tropics (Brown 1973). Genera such as *Formica, Lasius,* and *Myrmica* are found throughout the north temperate region (Wheeler 1910; Brown 1973). Some genera are common both in the tropical and temperate zones, e.g., *Aphaenogaster, Leptothorax,* and *Prenolepis,* and some individual species such as *Tetramorium caespitum* and *Lasius niger* disperse seeds at altitudes as great as 2300 m (Sernander 1906). In fact, the same author showed that ant dispersal occurs at both high altitudes and high latitudes and that no ant community he examined was so impoverished as to exclude seed-dispersing species. There is a general increase in ant-dispersed plant species from the poles to the equator that parallels the increasing diversity of ants across the same gradient. However, there may be a peak of abundance at the warm-temperate and "Mediterranean" midlatitudes (Beattie 1983).

The response of ants to elaiosome-bearing seeds and fruits can be extremely variable. Table 10 shows the behavior of ants toward violet seeds in four forest sites in West Virginia. *Aphaenogaster* was clearly the most efficient remover of seeds. A single ant species may remove the seeds of more than one plant species, but the frequency with which different seeds are removed differs according to the ant species (Table 11). Every study of ant dispersal has shown that each myrmecochorous species is serviced by a small array or assemblage of ants. For example, the seeds of *Viola odorata* are dispersed by three species of *Lasius,* two species of *Formica,* and one species of *Crematogaster* in central Europe. *L. niger* in the same region carries the seeds of twenty-six genera of myrmecochores (Sernander 1906). *Calathea microcephala* in Vera Cruz, Mexico, is dispersed by two species of *Odontomachus,* two species of *Pachycondyla,* several species of *Pheidole,* and a *Paratrechina* (Horvitz & Beattie 1980). *Sclerolaena diacantha* in northwestern New South Wales is dispersed by a species of *Iridomyrmex,* two species of *Rhytidoponera,* and one species of *Pheidole* (Davidson & Morton 1981b); and *Erythronium japonicum* is dispersed by at least two ant species in the forests of Honshu, Japan (Kawano et al. 1982). In every case, the seed-dispersing ant species represent only a small subset of the total number of ant species present in the habitat.

Table 10. *Numbers of different behavior patterns of ants toward violet seeds*

Species	Behavior				N
	"Ignore"	"Antennate" or "Examine"	"Attempt to pick up"	"Remove"	
*Aphaenogaster rudis**	19 (0.16)	20 (0.17)	6 (0.05)	71 (0.61)	116
Myrmica punctiventris	6 (0.15)	7 (0.18)	9 (0.22)	18 (0.45)	40
Lasius alienus	3 (0.11)	10 (0.37)	6 (0.22)	3 (0.30)	27
Tapinoma sessile	2 (0.20)	2 (0.20)	3 (0.30)	3 (0.30)	10
Leptothorax curvispinosus } *Lasius longispinosus* }	1 (0.25)	2 (0.50)	0 (0.00)	1 (0.25)	4
Formica subsericea	1 (0.20)	0 (0.00)	3 (0.60)	1 (0.20)	5
Stenamma schmitti	3 (0.50)	0 (0.00)	2 (0.33)	1 (0.17)	6
Formica integra	21 (0.52)	12 (0.30)	3 (0.08)	4 (0.10)	40
Myrmecina americana	5 (0.83)	1 (0.17)	0 (0.00)	0 (0.00)	6

Note: Each ant was counted only once and its strongest interaction recorded ("remove" is considered the strongest reaction and "ignore" the weakest); frequencies of each behavior are given in parentheses, and N is the total number of observations.

* *Aphaenogaster* spp. were lumped together because of difficulties in identifying individual foraging ants.

Table 11. *Frequencies of diaspores removed by various ant species in a West Virginia forest site*

Ant species	*Viola* sp. ($n = 98$)	*Hepatica acutiloba* ($n = 30$)	*Sanguinaria canadensis* ($n = 113$)
A. rudis	0.72	0.23	0.67
M. punctiventris	0.18	0.20	0.12
L. alienus	0.08	0	0.03
F. subsericea	0.02	0.57	0.18

Note: Frequencies are calculated so that for each plant species they sum to 1.

Variation in ant behavior toward seeds can seem almost idiosyncratic. *Aphaenogaster* and *Camponotus,* for example, have species that avidly disperse seeds in some areas, but not in others (Ridley 1930; Berg 1958, 1975; Culver & Beattie 1978). A species in a single habitat will disperse seeds one day but totally ignore them the next (Beattie & Culver 1981). This does not necessarily mean that the seeds are abandoned; they are often removed either by individuals from another nest, or of another species. Variation in the treatment of seeds by ants has occasionally been so great that authors working on the same plant species have reached different conclusions as to the importance of myrmecochory, for example, in *Allium* (Sernander 1906; Ernst 1979) and *Viola* (Culver & Beattie 1978; Schellner, Newell, & Solbrig 1982). However, de-emphasis of the ant–seed interaction has most likely been the result of field observations in habitats where prolonged disturbance has disrupted the ant community, or not carrying out the tedious and detailed experiments necessary to positively establish the presence of ant–seed interactions. The responses of any given ant colony to myrmecochorous seeds depend largely on the availability of alternative food and the nutritive content of elaiosomes (Culver & Beattie 1978).

Three types of ant–seed interactions have produced much debate as to whether the plants involved accrue a net benefit. The first is that of the harvester ants, most of which clearly exhibit adaptations such as powerful mandibles for the destruction of seeds. Many workers have long suspected that in some cases harvested seed is not eaten but is stored and abandoned at sites where germination and seedling establishment can occur (Ludwig 1899; Stager 1931; Uphof 1942; Davidson & Morton 1981b;

Andersen 1982; Heithuas 1983). Bullock (1974) and O'Dowd and Hay (1980), working with the harvesters *Pogonomyrmex* and *Veromessor*, respectively, have shown that this is possible, although the seeds concerned bore elaiosomes. However, two aspects of the seed–harvester interaction remain virtually unknown: first, the proportion of seed in storage that is eventually abandoned, and second, the probability that abandoned seed can germinate and produce a viable seedling. Specialized seed-eating ants may be restricted to habitats such as semiarid grasslands or deserts where large seed crops are produced predictably every year. For this reason they do not occur in myrmecochore-rich forests where most of the elaiosome-producing plants are small, vegetatively propagating perennials that rarely produce large seed crops and frequently produce none at all.

A second area for debate was stimulated by the finding that the seeds of *Calathea* in southern Mexico were taken, in large part, by ants known previously only as carnivores. Both genera, *Pachycondyla* and *Odontomachus* possess large slicing mandibles and powerful stings. In response to doubts that such obvious carnivores could disperse seeds, Horvitz (1981) showed through meticulous observation that these aggressive predators carry seeds back to the nest, feed the aril to the young, and bury the seed in the waste dump. This behavior lends credence to the theory that some elaiosomes mimic the prey of ants biochemically (Carroll & Janzen 1973). The plants clearly benefit as seeds treated in this way are more likely to give rise to new plants than untreated seeds (Horvitz & Beattie 1980; Horvitz 1981).

A third, debate-provoking, category of responses to seeds is that of *Formica* species that construct large mound nests. In his landmark monograph on seed-disperal by ants, Sernander (1906) reported that *F. rufa*, which constructs huge mounds up to a meter and a half in height, is a major seed disperser in various parts of Europe. He attributed the patterns of vegetation, especially the distribution of ant-dispersed species, to the activities of these very conspicuous ants. Ulbrich (1939) reinforced this conclusion and published a map of ant-dispersed species around some mounds. Unfortunately, the map revealed only a very loose association of these plants with nests, and there are no data at all on the patterns of distribution away from the mounds. Culver and Beattie (unpublished data) analyzed the vegetation around *F. rufa* mounds in southern England and found no consistent pattern of ant-dispersed species. In one case their density was positively correlated with distance from mounds, in another negatively correlated. In North America, Beattie and Culver (1977) and Culver and Beattie (1983) have shown that ant-dispersed species avoid the vicinity of mound nests; that is, given an apparently uniform habitat, ant-dispersed species are more abundant and diverse the farther they are from mounds. Other types of mounds such as the turf nests of *L. flavus* also show no particular association with ant-dispersed species (Dymes 1916; King 1977a). Large mounds may house colonies that are both too active and too permanent for ant-dispersed seed. The continual heaping of soil and debris may bury them too deeply, or the nest cleaners may chew the seedlings. However, it is possible that dormancy in some species is sufficient to outlast some kinds of nests and that emergence from an abandoned mound can occur. It should be emphasized that some species of *Formica* that build small ephemeral nests are bona fide seed dispersers (Culver & Beattie 1980), and some large mound nests, such as those of *Rhytidoponera* in Australia, are optimal sites for ant-dispersed plants (Davidson & Morton 1981a, b); this depends, however, on the ant species involved (Drake 1981). Some ants, for example, bring nutrient-deficient soil from deep layers to the surface, making the mound inhospitable to many plants (Briese 1974).

In summary, the selective advantage of relocation to nests and waste dumps is probably multiple. In some Australian vegetation subject to periodic fires and growing on impoverished soil, escape from fire and nutrient enrichment of microsites are probably at a premium, and escape from predation perhaps less crucial (Milewski & Bond 1982). In other habitats such as the eastern forests of the United States, where seed predation by rodents is intense and nutrients are monopolized by the dominant trees, escape from predation and nutrient enrichment may be the key advantages offered by ant nests. Add to this a couple of sympatric species in the same genus with and without elaiosomes, and avoidance of seedling competition becomes a third advantage. These examples illustrate the complexity of this class of ant–plant mutualisms and point to the large amount of careful experimentation that still needs to be done.

7

Ant pollination

In a world flora that harbors pollinators as diverse as slugs, mosquitoes, honey-possums, hummingbirds, and thrips; involving mechanisms as bizarre as pseudocopulation, pseudoaggression, and floral fermentation; and with reproductive structures as simple and ephemeral as the buttercup or as complex and long-lived as the *Banksia* inflorescence, it is very strange indeed that ants have not played a greater part. There are very few well-documented cases of pollination by ants. On the contrary, ants are widely regarded as thieves, parasitizing plants by taking floral rewards intended for pollinators, without performing the movements necessary for pollination (McDade & Kinsman 1980; Wyatt 1981; Fritz & Morse 1981; Willmer & Corbet 1981; Schaffer et al. 1983), or by simply chewing floral organs such as the style and ovary (Galen 1983).

Ant pollination has been reported a number of times: *Herniaria ciliolata* (Proctor & Yeo 1973), *Orthocarpus pusillus* (Kincaid 1963), *Polygonum cascadense* (Hickman 1974), *Glaux maritima* (Dahl & Hadac 1940), *Seseli libanotis* (Hagerup 1943), *Morinda royoc, Cordia brownei* (Percival 1974), *Rohdea japonica* (Migliorato 1910; but disputed by van der Pijl 1955), and *Microtis parviflora* (Armstrong 1979). *Diamorpha smallii* (Crassulaceae) was studied by Wyatt (1981) and Wyatt and Stoneburner (1981), who showed that this diminutive plant is also pollinated by ants, especially *Formica shaufussi* and *F. subsericea.* Pollen adheres to the hairs and integumental sculpturing of these ant species. They visit the flowers systematically but the degree of dependence on ant-borne pollen for seed set remains unknown. Petersen (1977) showed that the pollen of *Eritrichium aretioides* (Boraginaceae), *Oreoxis alpina* (Umbelliferae), and *Thlaspi alpestre* (Cruciferae) was carried by *F. neorufibarbis* in the alpine tundra of Colorado, and concluded that all three species are pollinated by ants. Various species of *Euphorbia* (Euphorbiaceae) and *Suaeda* (Chenopodiaceae) have also been reported as ant pollinated (Ehrenfeld 1978; Blackwell & Powell 1981). Hagerup (1943) suggested that ant-pollinated species may be relatively more abundant in some types of desert. The basis for this was his observation that in some kinds of low-growing vegetation in arid environments winged pollinators such as bees seemed scarce, but ants were exceedingly abundant. However, abundance is not necessarily correlated with pollinator effectiveness.

Hickman (1974) suggested a "syndrome" of plant characters that together indicate adaptation to pollination by ants. These characters were derived from a study of *P. cascadense* in which he demonstrated pollination by the ant *F. argentea.* Hickman predicted that most ant-pollinated plants would be short or prostrate, and densely growing or matted. These characters would ensure that ants were able to visit many individual plants without having to return to the ground. The flowers would be small and sessile, and the floral rewards minimal, so that each flower would attract a foraging ant combing the substrate, while remaining unattractive to winged pollinators searching for the kind of conspicuous floral displays that indicate the presence of rewards sufficient to maintain the energy demands of flight. Nectar would have to be easily accessible as ants have short tongues, and Hickman further suggested that there would be an advantage if the amount of pollen per flower were small. This might reduce grooming by the ant and hence the amount of pollen removed by the pollinator itself. Finally, ant-pollinated plants would be most abundant in hot, dry habitats where ant activity is high.

As is the case with other pollination syndromes a great many exceptions can be found, and Hickman's original suggestion requires some reinterpretation. A few orchids are reported to be ant pollinated: *Microtis parviflora* (Armstrong 1979), *Leporella fimbriata* (Bates 1979), and *Epipactis palustris* (Brantjes 1981). These are all "ground" orchids (i.e., terrestrial rather than epiphytic) with flowers borne on a single stem. Although none of them are tall plants, they are not prostrate and do not form a dense mat of branches. Their flowers are conspicuous to flying insects and pollen is deposited as pollinia. These traits do not fit the ant-pollination syndrome described by Hickman. Other plant traits such as height are irrelevant to arboreal ant species, which pass easily from branch to branch and from tree to tree high above the ground. Similarly, ants foraging in the herbaceous layer pass from plant to plant via overlapping foliage without descending to the ground. The requirement of a hot, dry habitat is clearly not true for ant-pollinated alpine plants. These habitats are also exceptionally poor in ant species (Petersen 1977).

There remain two more fundamental problems with ant-pollination research and the interpretation of the data. First, field experiments in which supposed ant-pollinated plants have been manipulated to demonstrate the necessity or efficacy of ant visits for seed set have rarely been carried out. Numerous observations have been made of ants visiting flowers, but almost nothing is known about their contribution to pollen transfer and subsequent seed set. The second problem arises directly from the first. In many reports of ant pollination, other floral visitors are reported as being numerous. Most conspicuous among these are small solitary bees from genera such as *Andrena* and *Dialictus,* and syrphid flies (hoverflies). On reflection, in the absence of crucial field experiments such as those outlined earlier, Hickman's syndrome could be interpreted as adaptation to small flying insects rather than ants. The emphasis of this syndrome is the miniaturization of plant and floral characters to the scale of ants, which are presumed to be tiny. However, many solitary bee species are smaller than *F. subsericea,* which pollinates *Diamorpha,* and forage over similar or even more restricted areas. With the lack of experimentation and data it is difficult to distinguish between ants and small-to-tiny flying insects as the principal, possibly coevolved, pollinators of most of the plant species listed at the beginning of this chapter.

Both botanists and entomologists have speculated as to why ant pollination is not more common. Botanists have inferred from the anatomy and morphology of flowers and inflorescences that ants have a negative effect on pollination and point to various devices that limit access by ants. These include physical barriers such as sticky tissues and glandular hairs, nectars and floral organs containing repellent chemicals, and the presence of extrafloral nectaries, which divert ants from flowers (Kerner 1878; Stager 1931; van der Pijl 1954; Bentley 1977a; Janzen 1977; Baker & Baker 1978; Feinsinger & Swarm 1978; Schubart & Anderson 1978; Rico-Gray 1980; Stephenson 1981; Elias 1983; Kevan & Baker 1983). However, these are only proximate answers. After all, why should these mechanisms have evolved in the first place? Why should plants prevent ants from reaching the flowers?

Entomologists have come up with quite a different set of reasons why ants would make poor pollinators. First, it has been argued that pollen cannot adhere to the integument of ants. In reality, however, many ant species are as hairy as many bee and wasp species, or even hairier. If ants are not hairy or bristly, the integument is often deeply sculptured and pollen is readily trapped in the furrows and grooves (Cole 1940; Sparks 1941; Hickman 1974; Wyatt 1981). Second, the objection has been raised that ants groom their bodies too frequently to be effective pollen vectors. Ants do groom themselves frequently, but then so do many important pollinators such as bees. In fact, some bees groom while in flight between flower visits. They remain effective pollinators because the stigma receives pollen from a part of the integument that is less accessible to grooming (e.g., Beattie 1971). Moreover, once stored in the "pollen baskets" of bees, pollen loses most of its viability (Kraai 1962; Heinrich 1979; Iwanami et al. 1979). The hazards of grooming may also be avoided to some extent by limiting the size of pollen loads as Hickman (1974) suggested, or by gluing pollen to the integument in various ways as in the Orchidaceae and Asclepiadaceae. It is also quite clear that these hazards are a routine component of numerous pollination systems in which pollen is deposited on pollinators in sufficient quantities to saturate grooming mechanisms. A third reason why some believe ants cannot be pollinators is that they do not fly and consequently cannot forage across distances that promote gene flow among plants. Many recent pollination studies suggest that this argument is incorrect. In the first place, many ant species forage over far greater distances than winged pollinators that are limited by small body size, the defense of a small territory, or that meet their nutrient and energy demands by foraging in highly localized, densely packed floral resources (Elton 1932; Sudd 1967; Wilson 1971; Linhart 1973; Davidson & Morton 1981b). With respect to gene flow, it is already thoroughly established that pollen transport in a great variety of pollination systems is severely limited and that most pollen is taken only a short distance from the parent plant. In many cases this results from the leptokurtic distribution of pollinator flight distances so that most pollen travels only one or two meters (Levin & Kerster 1974). These systems often result in small effective population sizes, which may be of great selective advantage (Wright 1969). The frequency distribution of ant-foraging distances may be leptokurtic (Wyatt & Stoneburner 1981) or platykurtic (Beattie & Culver 1979) and is therefore unlikely to differ greatly from those of important winged pollinators like bees and wasps. As a corollary to this particular argument it should be remembered that gene flow can also be accomplished effectively by fruit and seed dispersal, rather than by pollen transport (Levin & Kerster 1974). In addition, a low growth form or high density of plants may mean that as many individuals are accessible to ants as they are to winged pollinators (Hickman 1974; Wyatt 1981). A fourth and widely held argument against ants as pollinators is that ant foraging is neither systematic nor selective enough to service pollination mechanisms. In fact, ants have sophisticated sensory and orientation systems and systematically visit plants of all sizes, from low herbs to tall trees, to harvest resources as diverse as insect prey, honeydew, extrafloral nectar, and seeds (Berg 1954; Brian 1955; Tevis 1958; Wilson 1971; Laine & Niemela 1980; Lu & Mesler 1981).

In view of the refutations of these arguments it is appropriate to wonder why ants have not coevolved with flowering plants as pollinators in common with other social Hymenoptera. The fossil record shows that ants were present during the explosive adaptive radiation of the angiosperms during the late Cretaceous (see Chapter 2). By contrast, there are as yet no fossil remains of bees or wasps for this period (Wilson et al. 1967; Carpenter 1977). Although many paleontologists believe that bees and wasps are also of this antiquity, it is certain that ants were extant during the era when flowers and their attendant pollen vectors were radiating into a myriad of diverse pollination systems. It is true that one can point to many contemporary ant species that exhibit behavioral repertoires inappropriate to pollination, but there appears to be no compelling reason why mutual adaptations between early angiosperms and ants could not have led to the evolution of pollination systems as efficient and effective as those exhibited by contemporary bees and wasps.

A more fundamental answer to the problem of the rarity of ant pollination may lie in the chemistry of the interaction. Ants have many types of glands that secrete a wide variety of compounds. The bull ant *Myrmecia gulosa,* for example, has nine different kinds of exocrine glands involved in attack and defense, chemical recognition, and social organization (Cavill & Robertson 1965). The glandular secretions of other ants such as the leaf cutters may be even more complex (Weber 1966; Cherrett 1972), and the regulation of their fungus cultures involves, in addition, compounds present in the fecal fluid (Martin et al. 1975).

Schildknecht and Koob (1970, 1971) and Schildknecht (1976) have isolated β-indoleacetic acid (IAA), phenylacetic acid (PAA), and β-hydroxydecanoic acid (myrmicacin) from the metapleural glands of the ants *Atta sexdens* and *Myrmica laevinodis,* PAA and myrmicacin from *Messor barbarus,* and IAA and myrmicacin from *Acromyrmex subterraneus.* IAA and PAA are plant auxins. The complex secretions are antibiotic, inhibiting the growth of bacteria such as *Escherichia coli* and *Staphylococcus aureus,* and fungi such as *Botrytis, Alternaria,* and *Penicillium* (Maschwitz, Koob, & Schildknecht 1970; Maschwitz 1974; Schildknecht 1976).

In a series of experiments the effects of the components of the secretions of the metapleural gland on pollen grains have been elucidated. The effect of IAA on pollen function is extremely variable, depending on the species being examined, the concentration, and the presence or absence of other hormones. For example, Shukla and Tewari (1974) found that pollen-tube growth in *Calotropis procera* was promoted by IAA alone, but was inhibited when IAA was applied with small quantities of other plant hormones. Kwan, Hamson, and Campbell (1969) and Bhandal and Malik (1980) showed that IAA at very low concentrations increased germination and tube growth, but both were inhibited as concentrations were raised. In other species Mehan and Malik (1975) found that IAA was a pollen-tube inhibitor even at very low concentrations.

The effect of myrmicacin on pollen function is more consistent and acts, in general, to inhibit germination, retard pollen-tube growth, and disrupt pollen-tube mitosis (Iwanami & Iwadare 1978, 1979; Iwanami et al. 1979, 1981; Nakamura, Miki-Hiroshige, & Iwanami 1982). These authors, working with the pollen of a variety of plant species, found that at concentrations as low as 10 ppm, pollen germination was reduced, and among the pollen tubes that did emerge, the flow of materials for cell-wall synthesis, the function of Golgi vesicles, and pollen-tube mitosis were disrupted.

These effects were all detected in laboratory cultures. In collaboration with my colleagues Christine Turnbull, Bruce Knox, and Elizabeth Williams, I explored the possibility that pollen exposed to living ants would experience similar dysfunctions. It can be seen from Table 12 that twelve species were used, representing ten genera and six subfamilies. We used four species of pollen: *Prunus avium, Lycopersicon peruvianum, Rhododendron arboreum,* and *Acacia retinodes*; the experimental procedures are described in Beattie et al. (1984).

The results of these experiments were clear. Percent germination, percent viability, and pollen-tube lengths were significantly reduced by ant treatment in every case (Tables 12 and 13).

In order to see if the inhibitory effects of ants on pollen were reflected by diminished seed set, we pollinated flowers of the wild tomato (*L. peruvianum*) with pollen that had been exposed to ants. In this case it was unattached pollen taken from vials that had contained ants for thirty minutes. In the first experiment, twenty-three flowers were pollinated with pollen exposed to *Aphaenogaster* sp. and there were twenty control flowers. There was a significant reduction in seed set:

Control: $\bar{X} = 71.55 \pm 5.92$ seeds per fruit
Aphaenogaster: $\bar{X} = 51.13 \pm 5.79$ seeds per fruit
$P = 0.02$; $t = 2.45$; $df = 41$.

In a second experiment twenty-one flowers were pollinated with pollen exposed to *Myrmecia* sp., with twenty control flowers. Again there was a significant reduction in seed set:

Control: $\bar{X} = 71.55 \pm 5.92$ seeds per fruit
Myrmecia: $\bar{X} = 40.48 \pm 5.43$ seeds per fruit
$P = 0.001$; $t = 3.87$; $df = 39$.

Table 12. A, *Percent germination of control and experimental (ant-treated) pollen;* B, *Percent viability of control and experimental pollen*

Ant		Control		Experimental		
Subfamily, genus, and species		%	*n*	%	*n*	t_s
A. Percent germination						
				(*Prunus avium* pollen)		
Nothomyrmeciinae						
Nothomyrmecia macrops	(1)	75	800	56	465	11.55
	(2)	57	800	43	200	3.57
	*	75	800	55	240	12.13
	**	75	800	43	400	19.02
Myrmicinae						
Aphaenogaster sp.	(1)	75	200	47	200	5.84
	(2)	48	512	17	701	8.79
	(3)	67	801	47	801	8.33
	(4)***	61	801	33	800	11.38
	(5)****	60	800	34	800	10.54
	**	67	801	33	800	13.88
Chelaner sp.		64	600	26	802	14.32
Meranoplus sp.		64	600	26	800	14.32
	**	64	600	18	798	17.85
Pheidole sp.		67	801	44	800	9.35
	**	67	801	31	800	14.74
Ponerinae						
Amblyopone australis		67	801	58	801	3.73
Rhytidoponera sp. 1		48	512	30	800	6.57
sp. 2		42	600	32	800	3.82
Myrmeciinae						
Myrmecia pilosula		64	600	8	628	22.31
sp. 2		65	800	51	400	4.65
Formicinae						
Oecophylla smaragdina		59	600	39	500	6.63
Dolichoderinae						
Iridomyrmex sp.		65	800	44	763	8.40
				(*Rhododendron arboreum* pollen)		
Myrmicinae						
Aphaenogaster sp.		48	512	17	701	11.68
Ponerinae						
Rhytidoponera sp. 1		48	512	30	800	6.52
Formicinae						
O. smaragdina		71	800	53	800	7.62
B. Percent viability						
				(*Lycopersicon peruvianum* pollen)		
Aphaenogaster sp.		92	1506	63	780	16.43
*M. pilosula**		92	1506	78	860	9.24

Table 12 *(cont.)*

Ant		Control		Experimental		
Subfamily, genus, and species		%	n	%	n	t_s
			(*Acacia retinodes* pollen)			
Aphaenogaster sp.*		91	700	46	20	4.59
M. pilosula	(1)*	91	700	14	32	9.78
	(2)*	63	375	28	131	7.08

Note: Except where indicated, all treatments are for pollen removed directly from the ant integument. In each trial four ants were used, yielding four culture slides per trial and a total of 120 ants. The term t_s is a test of equality of control and experimental percentages (Sokal & Rohlf 1981). In all cases the value for t_s is highly significant, $P < 0.001$, showing differences between control and experimental percentages. Numbers in parentheses indicate replicates using fresh batches of pollen. Identification of ants to species level was impossible in some cases.

* Experimental values were obtained by allowing ant to walk undisturbed over culture slide.

** Experimental values were obtained from unattached pollen.

*** Values were obtained over 12 hours' incubation.

**** Values were obtained over 36 hours' incubation.

Table 13. *Statistics for pollen-tube lengths of experimental and control pollen*

Treatment	$\bar{x} \pm$ SE	n	t	P
1. Control	102.96 ± 1.90			
Aphaenogaster sp.	93.73 ± 2.07	800	3.29	< 0.01
2. Control	102.96 ± 1.90			
Pheidole sp.	93.90 ± 2.02	800	3.28	< 0.01
3. Control	87.33 ± 3.22			
Meranoplus sp.	51.22 ± 1.72	400	9.90	< 0.001
4. Control	87.33 ± 3.22			
Meranoplus sp. (unattached pollen)	53.34 ± 1.82	400	9.20	< 0.001
5. *Meranoplus* sp. (whole ant)*	51.22 ± 1.72			
Meranoplus sp. (unattached pollen)	53.54 ± 1.82	400	0.85	n.s.

Note: The data are shown as raw units taken from an eyepiece graticule. The conversion factor is 1 eyepiece unit = 0.0125 mm.

* To determine if the different methods of transferring pollen to culture slides after ant treatment yielded different results, the two *Meranoplus* treatments were compared.

Table 14. *Comparisons of percent germinability of untreated* Acacia retinodes *and* Brassica *pollen with pollen treated with metapleural secretion, using the G test*

Test	Pollen germinability (%)			
	Secretion from metapleural gland	Control	*G*	*P*
A. retinodes				
Ant 1	42(203)	51(227)	15.91	<0.001
2	24 (33)	44(282)	16.40	<0.001
3	58 (67)	83(117)	5.00	<0.05
4	21 (42)	41(123)	5.67*	<0.025
5	45 (47)	66(116)	5.88*	<0.025
6	10(218)	83(117)	180.65	<0.001
7	57 (51)	92(200)	30.76	<0.001
8	10(138)	92(200)	252.32	<0.001
Brassica				
Ant 1	23(335)	81(338)	245.08	<0.001
2	23 (88)	95(190)	161.29	<0.001
3	7(198)	99(203)	421.67	<0.001
4	4(200)	83(200)	294.53	<0.001

Note: $df = 1$ in each case. Sample sizes are shown in parentheses.
* The Williams correction was used because of small sample size.

Further experiments in which pollen of *Brassica* and *Acacia* were treated with secretion pipetted from the metapleural gland of the ant *Myrmecia forficata* showed that substances from this gland seriously damage pollen (Table 14). However, the presence of this gland is not crucial as exposure of pollen to the integument of a large worker of *Camponotus,* which has no metapleural openings, also severely disrupts pollen function (see Table 15; Beattie et al. in press).

It is clear from these data that exposure to living ants disrupts pollen function. It is not clear what causes this disruption or why ants exhibit these effects. As to the cause, the mechanism of action of myrmicacin makes it a strong candidate for pollen disruption. In the lab it has been shown to interrupt the flow of materials for cell-wall synthesis, interfere with Golgi vesicles, and disrupt pollen-tube mitosis (Nakamura et al. 1982). Thus it is potentially capable of affecting both germination and

Table 15. *Comparisons of percent germinability of ant-borne* Brassica *pollen with control pollen that had not been exposed to ants, using the G test (Sokal and Rohlf 1981)*

	Germinability (%)			
Ant species	Ant-borne pollen	Control pollen	*G*	*P*
M. forficata				
Ant 1	66	81	20.68	< 0.001
2	72	93	71.92	< 0.001
3	67	84	26.16	< 0.001
4	78	90	17.08	< 0.001
5	78	87	11.11	< 0.001
Camponotus sp.				
Ant 1	4	93	668.43	< 0.001
2	78	83	29.55	< 0.001
3	10	92	609.81	< 0.001
4	7	92	632.12	< 0.001

Note: $df = 1$ in each case.

pollen-tube growth. However, ants bear many glands with many products and so although secretions from the metapleural gland of one ant species are potent pollen antagonists, neither they nor myrmicacin can be thought of as the only source of pollen disruption until more experimentation has been done. At the same time it should be emphasized that although myrmicacin has been isolated from only four species of ant, they represent a varied sample, both in geographical and biological terms. *Atta sexdens* and *Acromyrmex subterraneus* are fungus-growing ants from the neotropics. *Messor barbarus* is a notorious seed harvester that is widely distributed around the Mediterranean Sea, both in Europe and North Africa. The fourth species, *Myrmica laevinodis,* is much less of a specialist than the other three, being omnivorous. It is widespread throughout northern Europe and adjacent Asia.

The possibility that ants may secrete compounds inimical to pollen function is not surprising when it is recalled that many kinds of animals affect plants chemically. Plant growth regulators of animal origin are quite common (Miles 1968). Many plant-sucking Hemiptera synthesize IAA from endogenous metabolites; that is, they do not merely recycle it from plant food, and their salivary products affect the growth of the host plant (Miles & Lloyd, 1967; Hori, 1976). Plant growth regulators are also suspected in the salivary secretion of many orthopterans (Dyer & Bokhari 1976; Detling & Dyer 1981) and a variety of large mammalian herbivores (McNaughton 1979).

Why should ants secrete compounds that disrupt pollen function? As long ago as 1910, Wheeler observed that ants produce an "oleaginous" substance that is spread over the body (Wheeler 1910). He surmised that this material inhibited the growth of bacteria and fungi. Some time later Bequaert (1922) wrote that ants were remarkably free of fungal attack relative to other insects. However, they are not completely immune (Figure 12). In recent years Maschwitz et al. (1970) and Maschwitz (1974) reported that ants distribute antibiotic secretions over the bodies of adults, larvae, and pupae. There has been some dispute over the glandular origin of these secretions but their existence, and the behavior leading to their dispersal over the body, appears to be beyond doubt. It is possible that these antibacterial and antifungal secretions incidentally inhibit the growth of pollen. If true, this would constitute an enormous counterselective pressure restricting the evolution of pollination by ants.

Seven species of fungi, commonly found in the soils where *Myrmecia forficata* nests, were tested for the effects of metapleural secretions (Table 16). Clearly the secretions are fungicidal or fungistatic, but the most intriguing results are for *Paecilomyces lilacinus* and *Beauveria bassiana.* These are entomophagous fungi, the former at least being a possible pathogen of ants. It is to be expected that an ant-borne antibiotic would be particularly effective against entomophagous microorganisms, and the data clearly show this.

The possibility that ants smear themselves with antibiotics may also explain why bees and wasps have such a different evolutionary history with respect to pollination. Ants differ fundamentally from bees and wasps in that their eggs, larvae, and pupae are deposited in natural cavities with no protection other than the constant attendance of workers. By contrast, the young of bees are deposited in wax cells of various kinds, and those of wasps in paper cells. Both the wax and the paper contain antibiotics, some of which are, in fact, also known to be harmful to pollen (Michener 1974). Consequently, ants are exceptional in that the juvenile

Figure 12 (*facing page*). An ant of the genus *Camponotus* that has been killed by the fungus *Cordyceps unilateralis.* The spore-producing structure has emerged from the head of the corpse. (From a specimen collected by Dr. Sophie Ducker.)

Table 16. *Results of exposure of seven species of fungi to metapleural secretions of the ant* Myrmecia forficata

	Experimental $\bar{x} \pm SE$*	Control $\bar{x} \pm SE$*	t	P
Paecilomyces lilacinus	4.4±1.29	131.2±9.12	−13.77	<0.001
Penicillium aurantiogriseum	5.8±1.53	98.4±14.62	− 6.29	<0.01
Trichoderma viride	17.2±0.97	71.8±5.68	− 9.47	<0.001
Gliocladium roseum	28.0±4.11	83.2±5.28	− 8.25	<0.0001
Beauveria bassiana	3.0±2.32	99.4±6.54	−13.89	<0.001
Aspergillus niger	23.6±3.09	62.2±1.93	−10.58	<0.0001
Cladosporium resinae	30.2±5.85	94.0±12.74	− 4.55	<0.01

* The mean and standard error for the number of hyphae traversing the cross-wires of a graticule under high power are given.

stages lie relatively exposed to attacks from microorganisms. Several authors have discussed the importance of microorganisms as a source of mortality in ant nests and have proposed that selection for defenses against them must have been critical ever since the origin of this method of brood rearing (Wheeler 1910; Evans 1974; Michener 1974). There is also a suggestion of this in the present study, as some of the most powerful effects on pollen were exhibited by the most primitive ant genera known: *Nothomyrmecia, Myrmecia,* and *Amblyopone.*

Although the habit of nesting in natural cavities without the construction of special brood cells confers the advantage of mobility in the event of danger to the colony, it also exposes the juvenile stages to bacterial and fungal attack. In the absence of brood cells, the smearing of every individual with antibiotics seems an extremely “logical” alternative. In the presence of secretions that inhibited microorganisms, and incidentally pollen, natural selection would yield on one hand an effective defense against pathogens, and on the other a powerful barrier to the evolution of ant pollination systems. Again it is tempting to implicate myrmicacin since it inhibits the growth of microorganisms and pollen grains and hence appears to be the common denominator. Whatever the compound(s) involved, this hypothesis, based on the chemistry of the interaction, has the merit of identifying a possible mechanism, and of offering an explanation as to why ants, unlike the other social Hymenoptera, are so unimportant as pollinators.

8

Food rewards for ant mutualists

Plants offer two basic types of rewards for the services of ants: housing and food. Housing was discussed in Chapters 2 and 3 and will not be taken up at length here. Food, or nutrition, will be treated in a broad sense. Thus, in addition to carbohydrates, proteins, and lipids ingested for normal metabolism, development, and growth, substances necessary for social organization such as those required for the biosynthesis of mating pheromones, alarm, and defense will be included. Also it should be borne in mind that substances that attract ants to a reward are not necessarily those from which they derive the primary benefit. That is, the distinction between attractants and nutrients may be crucial (Marshall, Beattie, & Bollenbacher 1979).

Ant nutrition

Most ant species involved in seed dispersal or plant protection are omnivores, and the plant rewards harvested by them while performing these services are only part of their overall nutrition. Individual colonies may opportunistically exploit extrafloral nectar, or elaiosomes on seeds at an intensive level for short periods; but almost invariably some workers are foraging at other resources at the same time. A major consequence of this is that the intensity with which plant rewards are harvested varies enormously.

This scenario is not at all surprising when it is remembered that the nutritional requirements of ants vary with the state of the colony, and conversely, the food available to a colony influences the number of individuals or castes. A few examples will illustrate this point. Foraging workers require carbohydrates, especially sugars, to provide energy for hunting, the subjugation of prey, the hauling of prey items or liquids back to the nest, and so on (e.g., Haskins & Haskins 1950). Honey solution made available to *Formica rufopratensis* colonies was fed preferentially to the larger workers (Gosswald & Kloft 1956); and Lange (1967), working with *F. polyctena,* showed that foraging workers received more honey solution than nurse workers that remained in the nest. An extreme situation was found in the same species by Schneider (1972) when foragers fed honey solution tended to retain it rather than pass it on down the trophallactic chain. An extremely careful study of sugar exchange in *Iridomyrmex humilis* by Markin (1970) showed that radioactively labeled sucrose solution was retained and utilized primarily by foragers, especially if the colony had access to the sugar up to a few hours before the experiment. On the other hand, a deprived colony rapidly shared the resource among the majority of the workers. Soldiers received a high proportion of sugars fed to colonies of *Solenopsis saevissima* (Vinson 1968).

Differences in sugar distribution within colonies vary according to species. For example, Wilson and Eisner (1957) compared colonies of several species fed honey labeled with radioactive iodide. Transmission of honey among individuals was almost absent in the seed harvester *Pogonomyrmex badius,* whereas the colonies of various *Formica* species were quickly saturated with the nutrient. The same researchers again confirmed that honey is transmitted chiefly among workers, with little being fed to the larvae or queens. For insects in general, a wide range of sugars is utilized although preferences for some have been demonstrated (Friend 1958).

With regard to proteins, Markin's (1970) study showed that labeled mosquito hemolymph was fed preferentially to larvae and queens, with relatively small amounts being retained by workers. The largest larvae received the protein first, although even the smallest were strongly radioactive after a few hours. The labeling of both sugars and proteins in these experiments clearly showed that larvae were fed primarily protein, and only small amounts of sugar. Queens also acquired protein more rapidly than sugar and were fed preferentially compared to workers. A similar preferential distribution of protein, first to larvae then to queens and finally to workers, has been shown by Vinson (1968) and Sorenson, Busch, and Vinson (1983). When deprived of a regular protein source the colony may suffer a variety of fates. Vanderplank (1960) fed nests of *Oecophylla longinoda* only honey solution and compared their growth with nests given a more complete, protein-rich diet, including *Drosophila* larvae and termite workers. After three months the honey-fed nests succeeded in producing only small, pale workers that hid when disturbed. By contrast, the nests with a complete diet produced normal, large, red-colored, aggressive workers. Stradling (1978) discussed several studies that showed that the production of reproductives requires an adequate protein intake by the colony.

Lipids of various classes are required for the normal metabolism of all insects and play a vital role in the function and organization of social insects such as ants. In particular, insects lack the sterol biosynthetic pathway and hence require a dietary or exogenous source of sterol, the only exception being those that harbor gut symbionts capable of performing this function (Clayton 1964; Gilbert 1967; Chippendale 1972; Svoboda et al. 1978). Linolate and linolenate are also dietary requirements for some insects (Chippendale 1972; Houk & Griffiths 1980) as is the lipogenic growth factor choline (Dadd 1973). Insect sterols have a variety of functions, being essential components of subcellular membranes, precursors of critical bioregulators such as molting hormones (ecdysones), components of surface wax of the cuticle, and constituents of lipoprotein carrier molecules (Steele 1976; Rockstein 1978).

Other lipids or lipoidal compounds perform some functions in ants. Wilson (1970, 1971) reviewed a variety involved in aggregating and alarm behavior in several ant species. The alkene undecane, from the Dufour's gland of *Acanthomyops claviger,* is a classic example of an aggregating pheromone. Lofquist and Bergstrom (1980) found a great variety of hydrocarbons in the Dufour's gland of *F. polyctena* that function as alarm, defense, and swarming substances. Law, Wilson, and McClosky (1965) showed that mixtures of terpenes, a class of derived lipids, are mating pheromones in *Lasius alienus* and *A. claviger,* and (z)-9-hexadecenal from *I. humilis* is both an aggregating factor and a trail pheromone (Cavill, Davies, & McDonald 1980; Van Vorhis Key & Baker 1982). Lipids have many other disparate functions: The metapleural gland of some ant species produces β-hydroxydecanoic acid, which may provide some basis for colony recognition (Brown 1968) or function as an antibiotic (Schildknecht 1976; Beattie et al. 1984) or both. The cuticle and wax layers of insects have high lipid fractions (Meinwald et al. 1975; Howard & Blomquist 1982), and juvenile hormones, being methyl esters of sesquiterpene epoxides, are lipoidal (Chippendale 1972).

Ants store lipids, occasionally in replete individuals (Glancey et al. 1973; Burgett & Young 1974). The storage of material in the lipid replete of *Myrmecocystus mexicanus* is mostly triglyceride with some sterol and fatty acids (Burgett & Young 1974). Triglycerides, either in repletes or in pharyngeal glands, are an extremely efficient method of storing energy. The complete oxidation of fatty acids yields about 9 kcal/g, whereas carbohydrates yield only about 4 kcal/g (Stryer 1981). The high surface volume ratio of ants suggests an additional advantage of fat storage. The oxidation of fats yields about twice as much metabolic water as carbohydrate (Dadd 1973; Downer & Matthews 1976). This may be of particular advantage in arid and semiarid environments. However, Ride (1970) described an incidental disadvantage for a species of *Iridomyrmex* that produces virgin queens in spring. They are 47.2% lipid and particularly attractive to spiny echidnas, which open nests specifically to find and eat them!

The pharyngeal glands of many ant species contain a store of lipids; for example, Barbier and Delage (1967) found large quantities of sterols, both cholesterol and various phytosterols, in the pharyngeal glands of the seed harvester *Messor capitatus,* and Peregrine, Mudd, and Cherrett (1973) showed that the pharyngeal glands of the leaf-cutting ant *Acromyrmex octospinosus* contained ten triglycerides in addition to several free fatty acids and sterols. Markin (1970) and Brian (1973) have demonstrated that the contents of these glands are fed primarily to larvae and queens. Vinson (1968) showed that soybean oil made available to colonies of *Solenopsis* was quickly harvested by foragers and fed preferentially to larvae and queens, with lesser quantities going to males, soldiers, and other workers.

There are few data on the precise nutritional requirements and biosynthetic pathways of ants. However, it may be inferred from studies of other insects that in general, a variety of sugars can be used as energy sources, primarily for active workers and soldiers. A variety of amino acids is required, especially for the growth of larvae, and lipids are crucial to development and metamorphosis with additional roles in colony organization and energy storage.

Elaiosomes

Bresinsky (1963) provided many insights into the nutritional rewards present in the elaiosomes of ant-dispersed plant species from northern and central Europe. Of the forty-one species analyzed all but three contained lipids and all but ten contained sugars. In addition, sixteen contained a significant amount of protein, nine contained starch, nineteen contained vitamin B_1 (thought to be a dietary requirement for some insects, Wigglesworth 1972), and twenty contained vitamin C. Glucose was the most common sugar (thirty species) and fructose (twenty-six species), saccharose (ten species), and xylose (two species) accounted for the remaining carbohydrates. Ricinoleic acid was a conspicuous component of the lipid fraction of eight species, as well as ant larvae. A variety of nutritional rewards was present in *V. odorata,* which contained lipids, including ricinoleic acid, sugars, proteins, and vitamins B_1 and C. Bresinsky concluded that ricinoleic acid was the ant attractant in this species. Marshall et al. (1979) analyzed the lipid fraction of the elaiosome of *V. odorata* in greater detail, taking advantage of a new bioassay that could be used both in the field and the lab. Exact quantities of fractions isolated from the elaiosomes were applied to porous Teflon cubes of about the same size, weight, and handling characteristics as the real violet diaspore. The cubes, being inert, permitted the bioassay of materials without contaminants, and their weight and texture permitted easy manipulation by ants. As the human experimenters also handled them easily, precise tests could be carried out in both the field and the lab.

During the preliminary experiments, the lipid fraction of the elaiosomes was clearly shown to be the most attractive to *Aphaenogaster* both in the field and the lab, and the nonpolar lipids were unequivocally more attractive than polar lipids. Gas-liquid chromatography revealed the presence of many fatty acids, and linoleic, stearic, palmitic, palmitoleic, nervonic, elaidic, vaccenic, lignoceric, myristic, behenic, 12-hydroxystearic, hydroxypalmitic, and ricinoleic acids were all bioassayed. Although ricinoleic, among other fatty acids, elicited some response from ants, it was not nearly as attractive as the diglyceride fraction. As the chromatography had revealed a large peak corresponding to oleic acid, it was possible that 1:2-diolein was the diglyceride attractant. In a final series of bioassays it was shown that 1:2-diolein was far more attractive than 1:3-diolein, monolein, oleic acid, and ricinoleic acid. Thus, contrary to Bresinsky's (1963) result, ricinoleic acid was neither a major fraction of *V. odorata* elaiosomes, nor did it elicit a strong response from ants. By contrast, diglyceride fractions and standards elicited the strongest response, in particular carrying behavior, which led to the removal of test cubes from the experimental arena – the behavior most relevant to seed dispersal.

The curious thing about the preceding experiments is that the ants preferred diglycerides, even to the constituent fatty acids. Diglycerides are known to be a major class of neutral lipid in the hemolymph of some insects (Gilbert & Chino 1974) and there is evidence that diglyceride-hemolymph protein conjugates are the means of lipid transport within insect bodies (Gilbert 1967). As they are also triglyceride percursors, it seems likely that ants respond to diglycerides as nutrients. Ant colonies fed fresh elaiosomes show greater survivorship than control colonies deprived of them (Steven Handel, personal communication). It should be noted that lipids are not always major components of elaiosomes; for example, O'Dowd (unpublished data) found that in eight Australian ant-dispersed species the lipid content of elaiosomes varied from 6.1% to 51.3% dry weight.

Lipids are important nutritionally but they also have other functions. Oleic acid, for example, elicits necrophoric behavior, that is, the removal of dead ants to the refuse pile or colony graveyard (Wilson, Durlach, & Roth 1958). This suggests that some lipid fraction of elaiosomes may function simply as a behavioral releaser. One interesting aspect of this possibility is the behavior of carnivorous ants such as species of *Odontomachus* and *Pachycondyla,* which are attracted to the arils (elaiosomes) of *Calathea* seeds in the rain forests of southern Mexico. These ants, armed with conspicuous slicing mandibles and powerful stings, are the major seed dispersers of some *Calathea* species (Horvitz & Beattie 1980). The possibility that elaiosomes somehow mimicked prey items, thus ensuring the removal of the seed to the ant nest, was first suggested by Carroll and Janzen (1973). Diglycerides, or closely related compounds, may be the common denominator since they occur in both hemolymph and elaiosomes. They may trigger the appropriate behavior (i.e., "carry this item back to the nest"), and elaiosomes may indeed mimic prey items biochemically.

The scant data on elaiosomes suggest two things: First, the complex array of substances in the tissue of some elaiosomes - carbohydrates, proteins, and lipids - may satisfy many of the nutritional demands of ants, including dietary requirements; and second, some substances in elaiosomes may function as behavior releasers instead of or in addition to being nutrients. Two types of behavior that may be triggered by elaiosomes are corpse carrying and prey carrying.

Extrafloral nectar

Various analyses of extrafloral nectars have shown that they contain sugars, amino acids, some lipids, and a variety of other organic compounds such as alkaloids and phenolics. The sugars have been reviewed by Bentley (1977a), who showed that sucrose, glucose, and fructose are the most common and that raffinose, arabinose, xylose, and rhamnose are occasionally present. The three primary sugars are found in the extrafloral nectars of many kinds of plants including a vine (Keeler 1977), a grass (Bowden 1970), a tree (O'Dowd 1979), several orchids (Baskin & Bliss 1969; Jeffrey et al. 1970), and a variety of ferns (Koptur et al. 1982).

These authors have also isolated small quantities of the sugars maltose, melibiose, gentiobiose, stachyose, lactose, and melezitose from various plant species, but some of these are not easily utilized by insects (Trager 1953). Clark and Lukefahr (1956) analyzed the extrafloral nectar of *Gossypium* sp. and then produced an artificial product based on the proportions identified in the analysis: sucrose, 40; glucose, 30; fructose, 20; raffinose, 5; rhamnose, 5; ribose, 2; and water, 100 units. Upon augmenting the diet of the pink bollworm moth with the artificial extrafloral nectar, they found a very significant increase in egg production and concluded that extrafloral nectar was important in maintaining fecundity.

The amino acids of extrafloral nectar have been reviewed at length by Baker, Opler, and Baker (1978). Twenty-two have been found with a mean of 14.70 ± 5.09 amino acids per nectar. Alanine, arginine, cysteine, glycine, isoleucine, proline, serine, threonine, and valine are particularly common, occurring in 80% or more of the species examined. Nonprotein amino acids occur more often in extrafloral nectar than in floral nectar. Baker et al. (1978) suggest that some of these may be toxic to certain potential visitors, perhaps reducing losses of extrafloral nectar to nonmutualist insects. The same authors were able to compare extrafloral nectars from vegetative organs and inflorescenses, but they found no significant difference in amino acid content between them.

Some extrafloral nectars provide the ten amino acids considered essential for the insect diet, but it is interesting that sulfur-containing amino acids of the cystine group are found more often in extrafloral than floral nectar. There is some evidence that insects in general require cystine for normal development, especially the formation of the pupa (Trager 1953). In laboratory tests, Lanza and Krauss (1983) showed that cystine was preferred to (more attractive than) several other individual amino acids. Amino acids such as glycine and tyrosine, although not normally considered essential, may also be dietary requirements for many insects but there are no data for ants. It is reasonable to suggest, however, that because extrafloral nectar does contain these particular amino acids, visitors such as ants may use it as a resource for the feeding of larvae.

Very little is known of the lipid content of extrafloral nectars. Herbert and Irene Baker (personal communication) have found lipids in fifteen out of eighty-five species examined, and suspect that they are, in part, free fatty acids. Keeler (1981a) found lipids in the postfloral nectar of *Mentzelia nuda* but did not report any further analysis. The paucity of lipid records may be merely the consequence of the difficulty and expense involved in the accurate identification of these compounds.

In their extensive studies of floral nectars, Baker and Baker (1975, 1983) showed that nectar constituents are functionally related to pollinator type. For example, flowers pollinated by hummingbirds frequently have more sucrose than hexose, and low concentrations of amino acids. By contrast, those pollinated by certain bees and butterflies generally produce hexose-rich nectar with high concentrations of amino acids. These differences reflect, among other things, the energy demands of different kinds of flight and the relative availability of alternative food sources for different pollinators. As Baker et al. (1978) pointed out, variation in the contents of extrafloral nectar may also reflect the differing requirements and preferences of visitors that provide a service for the plant. However, we can expect the patterns of variation in extrafloral nectars to differ from those in floral nectar for two reasons. First, access to floral nectars can be greatly restricted by floral morphologies that exclude nonmutualists. The long corolla tubes of hummingbird-pollinated flowers are well-known examples. By contrast, extrafloral nectaries are usually freely accessible to a wide variety of visitors, many of which are not mutualists. The secretion of chemical deterrents in extrafloral nectar appears to be one of the few antidotes, but this runs the risk of repelling mutualists with similar nutritional demands to those of unwanted visitors. Second, extrafloral nectar may provide rewards to mutualists with radically different requirements. For example, some extrafloral nectars rich in amino acids and lipids may be particularly attractive to ants requiring larval food. By contrast, the adults of parasitoid species visit extrafloral nectaries for their own immediate needs, which are primarily energetic. Energy sources are required for searching out hosts, and plant species adapted to parasitoid guards may secrete extrafloral nectars rich in sugars. Moreover, parasitoid larvae consume the bodies of their hosts and are consequently supplied with a ready-made and sufficiently balanced diet. The importance of extrafloral nectar in parasitoids has been shown by Lingren and Lukefahr (1977) working on the effects of the ichneumonid wasp *Campoletis sonorensis* on the cotton herbivore *Heliothis* sp. Wasps confined to varieties of cotton with extrafloral nectaries show increased longevity over those confined to nectariless cotton varieties with access only to floral nectar. Of greater interest is that although rates of parasitism of *Heliothis* were high on both cotton varieties, they were significantly greater on those with extrafloral nectar.

All these considerations lead to the suggestion that more detailed and comparative analyses of extrafloral nectars may reveal the following patterns: (1) A greater variety of nonnutrients, especially toxins and repellents, will be found in extrafloral nectar than in floral nectar; (2) plant species adapted to ant guards comprised mainly of omnivorous species will have a greater variety of nutrients, especially amino acids and lipids, especially if the plant species bearing the extrafloral nectaries appear at times when larvae or sexuals are a significant porportion of the ant colonies; (3) plant species adapted to ant guards consisting primarily of carnivorous species will have extrafloral nectar dominated by sugars (larval nutrients being supplied by prey hemolymph); (4) plant species adapted to a broad array of ant species, and parasitoids, will exhibit the greatest variety and concentration of nutrients; and (5) extrafloral nectar primarily for parasitoids will be particularly rich in sugars.

This picture may be complicated in at least two ways. First, extrafloral nectar may be involved in the attraction of pollinators in addition to, or instead of, its other roles (see Chapter 3). One would expect the constituents to reflect this. Second, extrafloral nectar may contain a variety of nonnutrient substances that are neither toxins nor repellents, but that attract parasitoids to hosts or to mates. Plant tissues contain a variety of molecules that are hymenopteran or dipteran parasitoid host-location cues or sex pheromones (Monteith 1967; Hendry et al. 1975, 1976; and references therein). One dramatic example, though not involving a parasitoid, illustrates this point well. Males of the oak leaf roller moth attempt to copulate with damaged host leaves exuding sap, in precisely the same way as they do with females (Hendry et al. 1975). If these highly specific molecules, which are frequently complex hydrocarbons or aliphatic acids, are secreted in extrafloral nectar, then the chemical, ecological, and evolutionary analysis of extrafloral nectar systems becomes an enormously more complex task than previously supposed.

Food bodies

As reported in Chapter 3, food bodies, including pearl bodies, are frequently food *tissues,* containing large amounts of protein and lipids in addition to the normal complement of cellular contents. The Beltian bodies of *A. cornigera* are vascularized and contain much protein, in addition to lipid (Rickson 1969, 1975). It is interesting that the extrafloral nectar of this species contains only five amino acids (Baker et al. 1978), suggesting a compensatory adjustment between the structures producing the ant rewards. The Müllerian bodies of *Cecropia* are remarkable in containing glycogen, a large branched polymer of glucose residues that is normally found in animal tissues. These food bodies contain 39% glycogen and 8% lipid by dry weight (Rickson 1971, 1973, 1976). This may be another example of a reward being a chemical mimic of prey items. The Beccarian bodies of *Macaranga* are especially rich in lipid (Rickson 1980) as are a wide variety of pearl bodies. O'Dowd (1982) has shown that the pearl bodies of thirty-six plant species all contain lipid, with a single exception: Those of *Ochroma pyramidale* contain sterols (O'Dowd 1980). Food bodies are a rich source of nutrients and are highly prized by some ants as larval food.

Rewards offered by homopterans and lycaenid larvae

Honeydew unquestionably serves as a reward for ants guarding those animals that secrete it (Way 1963; Atsatt 1981b). The honeydew of the ant-tended aphid *Myzus persicae* (Jones 1929) is among the most thoroughly analyzed and has been shown to be complex, containing sugars, amino acids, and amides (Auclair 1963) and lipids (Strong 1963, 1965). The dominant sugars are fructose, glucose, sucrose, and the oligosaccharides glucosucrose and maltotriosucrose. The honeydew of other homopterans on various host plants usually contains large amounts of the trisaccharide melezitose, and a variety of other oligiosaccharides. Melezitose is in fact characteristic of honeydew and is found in large quantities in many kinds of homopterans (Bacon & Dickinson 1957). Maltose is a common sugar constituent of coccid honeydew (Ewart & Metcalf 1956).

The eleven principal amino acids and amides found in *M. persicae* honeydew are alanine, asparagine, aspartic acid, glutamine, glutamic acid, leucine or isoleucine, phenylalanine, proline, serine, threonine, and valine. An identical list comes from analysis of coccid honeydew (Ewart & Metcalf 1956), and twelve more amino acids, including cystine, tyrosine, histidine, and tryptophan, have been isolated from various aphid, delphacid, and coccid honeydews (Gray 1952; Maltais & Auclair 1952; Auclair 1963; Sogawa 1982).

Lipids of various classes constitute between 12% and 16% dry weight of *M. persicae* honeydew, including free fatty acids, free sterols, with smaller amounts of triglycerides, and hydrocarbons such as squalene (Strong 1965). Twenty fatty acids have been identified, ranging from C_4 to C_{20} with palmitic acids (16:0) being the most abundant (Strong 1963). Linoleic acid (18:2) appeared in the phloem of the host plants of the aphids analyzed, but not in the honeydew, leading Strong (1963) to suggest that this acid may be an aphid dietary requirement.

Various honeydews have been shown to contain a variety of other molecules such as peptides, minerals, B vitamins, plant hormones, and hydroxy acids such as malic and citric (Gray 1952; Maltais & Auclair

1952; Way 1963). Many honeydew constituents appear to be derived almost unaltered from the ingested phloem juice. A conspicuous exception is melezitose, which is probably synthesized by gut symbionts. In fact, the presence of gut symbionts may make honeydew a particularly valuable resource for ants because they can provide dietary requirements such as sterols, which may be otherwise difficult to obtain.

Maschwitz et al. (1975) analyzed the honeydew secreted by the larvae of the butterfly *Lysandra hispana.* They found the sugars fructose, sucrose, trehalose, and glucose, a little protein, and the amino acid methionine. There was no search for lipids. The honeydew of an Australian lycaenid has been analyzed and shown to contain several more amino acids, notably serine (N. Pierce, personal communication).

The main conclusion from these data is that honeydew is a complex substance that provides many, and perhaps in some cases all, of the dietary requirements of ants.

Ant rewards: supply and demand

Existing data show that plant tissues and secretions that function as ant rewards, and honeydew, contain a wide variety of essential nutrients. Some of them, especially individual sugars and amino acids, are found at low concentrations and may not be very significant nutritionally to either the adults or larvae of a given ant colony. Nevertheless, the presence of sugars, amino acids, and lipids in most rewards suggests that plants "hedge their bets" by offering ants a variety of foods. Variation in supply and demand is a product of several groups of factors. The first group is derived from the age structure of the ant colony. In essence, the nutritional requirements of adults and larvae are very different. As we have already seen a young colony with few larvae requires energy sources, primarily sugars, to maintain its foragers, nurses, and soldiers. To be sure, the queen requires proteins and lipids for egg production, but not in the amounts required by cohorts of developing larvae. The more mature colony has an increasing demand for amino acids for growing larvae. Nonfeeding stages such as diapausing larvae and pupae require much more lipid as energy reserves. The production of sexuals, which is often late in the season, requires a larval diet rich in amino acids and possibly lipids. Reproductives are often produced in very large numbers to overcome the dangers of finding mates and locating new nest sites. Predation is a particularly important source of attrition at these stages as birds, lizards, toads, and other ant species feast on the relatively vulnerable sexuals. The production of sexual swarms is highly dependent on the colony's intake of protein-rich food.

Various seasonal effects complicate this picture, particularly the onset of drought or cold. Drought may be overcome, in part, by the storage of lipids, which are a rich source of metabolic water. Stored lipids, especially in larvae, may provide the energy and water required by a colony to last through extended dry or cold seasons, as the larvae of many ant species are known to regurgitate food when solicited by adults (Wilson 1971). Workers may also accumulate fat for overwintering (e.g., Kirchner 1964). Ant species subject to seasonal drought or cold, which halts or curtails foraging, are likely to exhibit a rise in lipid demand as the period of adversity approaches. Most of the demographic colony changes are characteristic of a particular season. This was highlighted by Banks and Macaulay (1967), who showed that *Lasius niger* assiduously tended aphids for honeydew while the sexual brood was developing, but rapidly lost interest after midsummer when the sexuals had matured and the high honeydew requirement had consequently declined.

A second group of factors affecting the nutrition of mutualistic ants is a product of spatial variation in ant activity. Perhaps first among these factors is the proximity of ant colonies to the plants requiring their services. The importance of proximity has been demonstrated by Laine and Niemela (1980), who found that the rate of *Formica* predation on defoliators of birch trees is largely a function of the location of the nest. Trees close to nests are hunted upon intensively whereas those more distant are more likely to lose their leaves. The net result is the formation of "green islands" around the nests. This showed that there are limits to the distance foragers can travel, determined by factors such as energetic costs, territoriality, and physical barriers.

Variation in alternative food sources is a second factor and is likely to be closely linked to the first. One can imagine many kinds of events that might divert foraging ants away from plants producing extrafloral nectar or elaiosome-bearing diaspores: the hatching of a cluster of caterpillars or the discovery of a batch of eggs or pupae on a plant closer to the nest; the maturation of other plant species that produce more attractive nectar or larger elaiosomes; the monopolization of plant rewards by a more aggressive colony or ant species and the consequent shifts in territory or foraging range; exposure to predators or dessication at the location of plant rewards; natural disturbances such as a tree falling or local flooding; catastrophes overtaking the colony such as fungal infection, invasion by nematodes, or a slave raid; and so on.

Table 17. *Number of occurrences of different behavior patterns of* Aphaenogaster *spp. toward chasmogamous and cleistogamous seeds in "cafeteria" experiments*

	Seeds tested in	
	June	July
"Ignore"	19(0.16)	7(0.22)
"Antennate" or "Examine"	20(0.17)	13(0.41)
"Attempt to pick up"	6(0.05)	3(0.09)
"Remove"	71(0.61)	9(0.28)
Total	116	32

Note: Only the "most interested" behavior of each is recorded; frequencies are given in parentheses. Removal frequency significantly different, $P < 0.01$

One likely example of events such as these appeared during a study of the seed dispersal of several ant-dispersed violet species in West Virginia (Culver & Beattie 1978). Five species produced chasmogamous seeds from cross-pollinated flowers in early summer, and cleistogamous seeds from closed, obligately self-pollinated flowers in mid and late summer. "Cafeteria" experiments were performed in which individual fresh seeds of each species were placed on the floor of relatively undisturbed forest and the behavior of ants toward them closely monitored. In June the seeds were quickly located by ants and taken back to their nests. During July and August the ants showed an increasing lack of interest in the seeds. For example, at a site called Big Draft the removal rate of chasmogamous seeds from experimental arenas in June was 1.55 seeds per hour, whereas the rate of removal of cleistogamous seeds in late July was 0.75 seeds per hour, a reduction of 50%. Differences in removal rates were apparently not the result of differences in the attractiveness of chasmogamous and cleistogamous seeds. In late July, when a few chasmogamous seeds were still available for testing against cleistogamous seeds, there was no difference in the frequency of removals when the ants were given a choice. Furthermore, cleistogamous seeds actually bore larger elaiosomes than chasmogamous seeds (Culver & Beattie 1978), an apparent "compensation" for the diminishing interest of ant dispersers illustrated in Table 17.

The seasonal difference in the rate of seed removals was partially the result of the greater availability of alternative foods in mid or late summer. It is unfortunate that no quantitative data were collected, but several observers watched the experimental arenas for many weeks on end and noted that the mandibles of ants approaching seeds in June were generally empty, whereas those of ants crossing the experimental arenas in late July and August were much more likely to be clasping a dead or struggling insect. It seems that elaiosomes are important food items until the insect populations build up; then the plants must compete with these for the ants' attention. A peak of interest in seeds on the part of ants in early summer may be the principal selection pressure that has led to early flowering and rapid fruit maturation in many ant-dispersed herbaceous species growing on the forest floor of northern deciduous forests.

Evidence for the constantly changing patterns of ant territories and foraging ranges comes from the work of Leston (1978) and others documenting the movements of colonies and species in the canopies of tropical forests. Vanderplank (1960) showed that nests may be abandoned as a result of the buildup of nest inquilines to intolerable levels; and Way (1963) reported the shifting interest of ants on plants supporting homopteran colonies, which are in turn subject to changes in density, viability, and productivity.

A third group of factors affecting the nutrition of mutualistic ants arises from variation in the availability of plant species offering rewards, and in their physiological condition. The density of extrafloral nectaries and amount of extrafloral nectar may be greater in a stand of small trees such as *Ochroma* than in a group of perennial herbs such as *Costus* or epiphytic orchids as a simple consequence of size. However, herbaceous perennials such as *Helianthella quinquinervis* that bear extrafloral nectars can reach high densities (Inouye & Taylor 1979), and it is thus difficult to compare the ant rewards of different plant species and communities. No field comparisons have been made to my knowledge. Nevertheless, local variation in the availability of plant rewards for ants is bound to reflect local variation in the composition of the plant species.

There is also variation in the rewards offered by any given plant species, both temporally within individuals, and among individuals. In Chapter 3 it was shown that the extrafloral nectaries of many species are active only for limited periods such as bud break and leaf expansion (e.g. Tilman 1978). Studies of cotton extrafloral nectar have shown that the volume of secretion is greatest from mid-July to mid-August, but that the sugar and amino acid content declined steadily through the season (Yokoyama 1978). In addition, Butler et al. (1972) demonstrated that *Gossypium barbadense* and *G. hirsutum* extrafloral nectar production decreased dramatically at night; but both the volume and concentration of sugars quickly rose to maximum levels at 8:00 a.m. and 11:00 a.m., respectively. O'Dowd (1979) found no differences in the diurnal and nocturnal secretion of extrafloral nectar in *Ochroma;* however, the volume of nocturnal secretion was dependent on the position of leaf insertion. His study showed that petiolar nectaries secreted for a mean of 17 ± 4 days and leaf-vein nectaries for a mean of 37 ± 3 days, and that ants were present during these periods. However, significant variability in the number of foraging ants per tree was found. O'Dowd attributed this in part to the fact that expanding leaves produced the most nectar and that the number of expanding leaves varied according to the individual tree. Those that grew at poor microsites or had been damaged produced fewer new leaves and were visited less by ants. By contrast, large, healthy trees sustained large ant populations and healthy foliage.

Variability in the volume and content of honeydew has been shown to depend on the plant species, the homopteran species, ambient temperature, age of the plant, and gut symbionts, among other factors (Auclair 1963; Llewellyn 1972).

Plant tissues, in general, have low concentrations of nitrogenous compounds, and variation in the levels of protein, in particular, profoundly affects the growth and fecundity of phytophagous insects (Russel 1947; McNeill & Southwood 1978). Other essential elements in plants, such as phosphorus, potassium, calcium, magnesium, sulfur, iron, and sodium, also vary according to plant species, growth form, soil type, and water availability, as well as the individual tissue under consideration (Woodwell et al. 1975; and references therein). The nitrogenous content of phloem and xylem sap varies according to similar factors (Mattson 1980). This affects phloem feeders such as aphids, which grow faster when amides are more abundant, in turn increasing the nitrogen content of honeydew (van Emden 1972). Variation in other ant rewards of plant origin such as extrafloral nectar and elaiosome tissue is little understood. Floral nectar is known to vary widely between species, and to change in volume and composition during the day and/or through the season of anthesis (Percival 1961; Baker & Baker 1975). Extrafloral nectar may be expected to vary in similar ways (Baker et al. 1978), and since many extrafloral nectaries and elaiosomes are vascularized (Frey-Wyssling 1955;

Bresinsky 1963), variation should be attributable, at least in part, to the changing composition of phloem and xylem fluids.

It is obvious that plants can exert some "quality control" over nectar content, thus facilitating the establishment of functional, mutualistic relationships with particular pollinators. Also, since phytophagous insects are both diverse and abundant, there are clearly strategies that circumvent the apparently bewildering variety in the nutritional contents of plant species and organs (Ahmad 1983). At the same time, however, it is against this kaleidoscopic chemical background of colony physiology, localized ecological conditions, and plant physiology that plants must attract and retain ant foragers for protection, seed dispersal, and plant feeding. In addition, the plant rewards such as extrafloral nectar, elaiosome tissue, and housing are often freely accessible to many kinds of nonmutualists, especially insects, which simply steal them. Therefore it is small wonder that ant protection and ant dispersal of diaspores break down even in the presence of rewards. The data discussed in this chapter suggest three ways in which rewards maximize ant services. First, the greater the array of rewards, the greater is the probability of ant service. This can be seen in those ant–plant mutualisms that appear to be most obligate. Thus, in the *Acacia–Pseudomyrmex* and *Macaranga–Crematogaster* plant protection systems, carbohydrates, proteins, and lipids are produced on a more or less year-round basis, generally meeting the nutritional requirements of the ants. Ant fidelity to the plants is high, or even obligate, and the probability that the plants will be occupied by aggressive species is also high. As we have seen in Chapter 3, the presence of domatia, extrafloral nectaries, and food bodies by no means guarantees that the plants will be occupied by efficient ant guards or that a considerable amount of herbivory will not take place. The deletion of any one or more of these rewards appears to diminish the protection afforded by ants (O'Dowd 1979; Andrade & Carauta 1982) and increase the importance of alternative means of defense against herbivores (Rehr, Feeny, & Janzen 1971; Coley 1983). This trend may be complicated by the presence of homopterans, which supply a broad array of nutritional requirements in honeydew and frequently occupy domatia, or flourish on plants independently of ant-reward structures.

The second way in which services may be maximized is the diversification of the nutrients and attractants in food rewards. In the obligate mutualisms where food bodies and extrafloral nectar are provided, the ants appear to receive a complete diet from food rewards. But even when there is only a single reward such as extrafloral nectar, honeydew, or elaiosomes, a variety of sugars, amino acids, lipids, and other nutrients or attractants is often available to ants.

Third, some rewards are obviously very potent; in other words, they are very strongly attractive to ants and the ants may have a great need for them, at least for brief periods. For example, earlier in this chapter a review of elaiosome constituents revealed that they often contain carbohydrates, proteins, and lipids. The analyses were mainly of ant-dispersed species from temperate deciduous forests that have been shown to (a) be subject to high rates of seed predation (Heithaus 1981); (b) produce small numbers of large seeds relative to non-ant-dispersed conspecifics, or other species in the same habitat (Salisbury 1942; Beattie 1983); and (c) propagate vegetatively (Beattie & Culver 1981). Each of these suggests that seed survivorship is low and seedling recruitment into populations an infrequent or sporadic event (Harper 1977). Given first, that the energetic and nutritional investment in the seeds is great, second, that the seeds nevertheless remain vulnerable, and third, that ants are the primary determinants of seed and seedling survivorship, a strategy to attract and maintain ant dispersers is likely to involve very potent rewards, in this case an array of essential nutrients in the elaiosome.

To my knowledge no elaiosome-bearing plants also produce domatia. However, ant dispersers of temperate forest plants are all ground nesting and there appears to be no shortage of nest sites (Headley 1952; Lyford 1963; Smallwood & Culver 1979). The potential importance of nest sites in the vicinity of ant-dependent plants is perhaps illustrated by *Viola cucullata,* one of seven violet species studied by Culver and Beattie (1978). Its seeds bear small elaiosomes, but because it grows in marshes and other waterlogged habitats where nesting is unlikely, they are rarely taken by ants. Some plant species with elaiosomes also possess extrafloral nectaries (Elias, Rozich, & Newcombe 1975; Horvitz & Beattie 1980), but there are no data on the possibility of positive feedback between the rewards increasing protection or seed dispersal.

In summary, it appears that some plants may overcome variations in supply and demand by offering an array of rewards to ants. Broad arrays increase the probability of ant service but do not guarantee it. They may generate an obligate relationship between ant and plant, the ant species being confined to the plant species that diverts large portions of its resources and structure to ant rewards. Plant species offering single rewards, or rewards present for brief periods only, show a decreased probability of ant service. However, even among single rewards, diversification of nutrients, dietary requirements, and attractants seems to be the rule. The possibility that extrafloral nectar or elaiosomes are adapted chemically to reward specific ant species, or suites of species with particular behavioral repertoires, is likely but so far undetected. Such specialization would increase the probability of ant service within the distribution of the target ants.

9

Variation and evolution of ant–plant mutualisms

The survey of ant–plant mutualisms in Chapters 3–7 presents a complicated picture. In Chapter 3 it was shown that careful studies of plant species that bear extrafloral nectaries have yielded contradictory results. Some show clear evidence of ant protection, others do not. In Chapter 5 the phenomenon of ant feeding of plants was confirmed in a couple of species, but the benefits conferred by ants on a great variety of others that harbor nests remained obscure. In Chapter 6 the discussion of ant dispersal showed that different authors can study the seed and seedling demography of the same elaiosome-bearing species and reach different conclusions as to the importance of ants at these life-history stages. The situation for ant pollination (Chapter 7) is also unclear. The data are scarce and claims of ant pollination, often based on anecdotal evidence, are generally inconclusive. Although it is possible that the differing conclusions reflect individual biases among authors, it seems more likely that they accurately reflect natural variation in function and effect. In this chapter the causes of this variation will be examined using ant protection and ant dispersal as examples.

The evolution of mutualisms is affected by demographic and life-history characteristics of the plant and ant populations. For the ants we know that particular foods are required at particular stages in colony development or reproduction. We also know that profound demographic changes such as the proliferation of worker castes or reproductives are likely to lead to increases and decreases in colony interest in plant rewards. Among plants there may be annuals with single brief bursts of new shoot or seed production, or perennials of various longevities and sizes that produce either a mass of new shoots for a limited period each year or a trickle of new shoots all year long, with seed production varying in both time and space. Each of these factors contributes to variation in the intensity of interaction.

Ecological factors affecting the evolution of ant–plant mutualisms include the abiotic setting and the other species with which the ants and plants interact directly and indirectly. These factors have occupied much of the other chapters and a couple of examples will suffice. An ant species that provides an effective service to a plant species with elaiosomes may be competitively displaced by another ant equally attracted to the reward but which leaves the seed at microsites totally inappropriate for germination or establishment (e.g., Turnbull 1984). Alternatively, the seeds of ant-dispersed species cease to be dispersed as the local ant community turns to more abundant food items (e.g., Culver & Beattie 1978).

Variation in the vulnerability of plant enemies to attack may affect the quality of protection. Overall there is a surprising paucity of data on what ants actually eat, attack, or repel while visiting plants with extrafloral nectaries. They do remove leaf and stem chewers such as dipteran, coleopteran, and lepidopteran larvae that live on or close to the plant surface (Chapter 3). On the other hand, there is little evidence that they have any effect on enemies such as root or stem bores, leaf miners, and many kinds of gall makers that succeed in becoming embedded in plant tissue. It appears that once a herbivore or seed predator burrows below the plant surface it becomes relatively inaccessible to ants, and other forms of defense become necessary. The effectiveness of ant guards is also limited by the distribution of extrafloral nectaries. Growing tips and expanding tissues such as apical and floral meristems and young leaves are the usual sites of nectar secretion and are especially attractive to ants. Old leaves, the undersides of leaves, bark, and branches close to the ground may be neglected by ant populations zealously guarding younger tissues elsewhere on the same plant.

The following lists summarize data on the demographic and ecological factors affecting the efficacy of ant services. The wording applies directly to the protection of plants but most entries apply in principle to seed dispersal by ants.

A. *Protection is enhanced when the local ant colonies or species*

1. Have a dietary/physical need for the reward(s) offered by the plants.
2. Have sufficient foragers to patrol the plants and harvest the rewards.
3. Have the appropriate behavioral repertoire to attack, disturb, or remove herbivores and/or seed predators.
4. Do not suffer from competition for the rewards with nonmutualists.
5. Nest near to or actually on the plants, including them in the normal foraging range.
6. Are not continually inhibited from foraging on the plants by bad weather, excessive shading, flooding, and so on.
7. Actively forage while the plants are most vulnerable.
8. Do not have a variety of attractive alternative resources available at the time when the plants require their services.

Three possible indirect factors:

9. Do not disrupt the pollination system by inhibiting or removing pollinators.
10. Do not disrupt the predators and parasites of plant enemies that the ants do not or cannot deter.
11. Do not tend herbivores such as homopterans to the extent that they reach destructive densities.

B. *Protection is enhanced when the plants*

1. Have a herbivore/predator problem that reduces fitness.
2. Offer reward(s) the ants require.
3. Offer sufficient rewards to maintain ant foragers on the foliage.
4. Provide rewards that are most accessible to and most efficiently harvested by ants, rather than nonmutualists.
5. Are located within the sphere of influence, or sphere of dominance, of a beneficial ant colony.
6. Are physiologically capable of allocating energy to maintain a flow of rewards during periods of vulnerability.
7. Are located where the supply of sunlight, water, and essential nutrients will not limit the production of rewards.
8. Are growing where the nests of mutualistic ants can be built, or the plants themselves provide nest sites.
9. Are not dependent on other insects such as pollinators or parasitoids that fall prey to the ants in large numbers.

Variation in the outcome of some other plant–insect mutualisms is very well documented. For example, individuals or populations of plant species dependent on insect pollinators frequently fail to set seed, or seed production is well below potential. Although this may be the result of many factors such as pollen incompatibility, a genetic trait, it often can be attributed to some failure in the flower–insect mutualism. In fact, numerous experiments have been carried out to test for pollinator limitation of seed set in which yields from undisturbed flowers open to their natural pollinators have been compared to yields from carefully hand-pollinated flowers (Bierzychudek 1981; Schmitt 1983). The latter yields are often significantly greater, which indicates the frequency with which the flower–pollinator mutualism works poorly and explains the wide variation in its effect (i.e. seed set occurring in natural populations).

Generalist versus specialist

Does selection favor the evolution of broad arrays of ant species that service any given plant species, or does it favor specialization? Benefits accrue to plants that attract a broad array of ants and in the case of extrafloral nectaries, an array of other potential guards, especially wasps. The greater the diversity of ants the greater the variety of plant enemies removed, and the greater the probability that in any given habitat, season, or time of day, some ant species will forage on the plants or take their seeds. On the other hand, the ecological, behavioral, and demographic characteristics of certain ant species clearly result in more effective services. If plant species exhibit variation in traits especially attractive to these ants, then there may be an advantage to restricting access to extrafloral nectar or elaiosomes to those ant species that provide the best services. Open rewards available to a broad array of ants appear to be a particularly appropriate strategy in the face of the constant changes in ant communities. Studies of a variety of habitats have demonstrated high turnover rates in species composition and major fluctuations in the abundance of individual ant species (e.g., Levins, Pressick, & Heatwole [1973]; Leston [1978]; and Baroni-Urbani, Josens, & Peakin [1978]). These are the result of many factors, not the least of which is the short nest life of many ant species (Smallwood 1982b). As a consequence, plants with ant rewards are not only subject to frequent shifts in arrays of ant species on a local scale, but expansion of their populations may place them outside the range of the most effective mutualists. These circumstances favor selection for general rewards. The great problem with this is that the rewards become available and attractive, not merely to less effective ant species, but also to a host of parasites and predators. This menagerie consumes extrafloral nectar and elaiosomes (often with the seeds) without any benefit to the plants (Chapters 3, 6). In fact, it seems that the load of parasites and predators (i.e., cheats and thieves; Thompson 1982) is a major determinant of the cost–benefit equation of the ant–plant mutualism. It is not difficult to imagine that if the load is too great the majority of the rewards are consumed by them, with the result that the service and the mutualism fail.

Selection and fitness

If ant–plant mutualisms evolve in this complex milieu of restraints it is little wonder that ant assemblages are in large part fortuitous, interaction is diffuse and facultative, and specialization between particular ant and plant species an exceptional occurrence. However, the countercurrents leading to the evolution and maintenance of ant–plant mutualisms are strong: O'Dowd and Hay (1980) suggested that the predictability of appropriate ant species is crucial; and it now seems clear that the legendary ubiquity and abundance of ants in most environments worldwide mean that a plant population that offers ant rewards will receive services from some local array of ants.

More important than this generalization is the demonstrable fact that the mutualistic component of fitness in ant–plant interactions is important: Individual plants with extrafloral nectar and ant guards suffer less grazing and can set more seed than those from which ants have been experimentally excluded (Chapter 3). Epiphytes grow faster and larger when associated with ant nests (Chapter 5). Seedlings emerging from ant nests are more numerous, larger, and longer lived than those denied ant manipulation (Chapter 6). Ant colonies profit from gathering ant rewards (Chapter 8). In other words, the evidence shows that major gains in fitness accrue to plants mutualistic with ants.

This is emphasized by a few studies that have explored the consequences for the plant when mutualism with ants breaks down. *Cecropia peltata* is an ant-guarded plant throughout lowland Central America and tropical South America. However, on Puerto Rico and many other Caribbean islands it fails to produce trichilia and Müllerian food bodies (Janzen 1973a; Rickson 1977), and does not attract ant guards. Janzen (1973a) has found a few plants on Puerto Rico with rudimentary trichilia, and another island off the coast of Nicaragua has plants in which they are moderately developed but are not full size. Clearly there is considerable genetic variation for these ant-related traits. In the wettest parts of the northeastern Puerto Rican lowlands, *C. peltata* does occasionally occur as saplings and as a few mature trees. By contrast with the mainland plants, however, these individuals are festooned with invading vines and very heavily grazed. Although it is true that the mainland *Cecropia* may be invaded, the level of overgrowth and herbivory on the island trees appears so great that few individuals survive for any length of time. *C. peltata* becomes far more abundant at higher elevations, even forming pure stands in places. Both competing vines and herbivores appear to be less common at higher altitudes. This suggests that ant guards determine in part the range and distribution of *C. peltata,* because when they are absent from habitats rich in vine and herbivore species, only scattered, struggling individuals can be found.

In his studies of *Acacia* in Central America, Janzen (1966) showed that ants were crucial to the survival of the plants. For example, 108 individuals of *A. cornigera,* which is normally occupied by *Pseudomyrmex ferruginea,* were followed for nearly a year. Ants were removed from 69 trees and the mortality of occupied and unoccupied individuals monitored.

After ten months 56.5% of the unoccupied trees and 28.2% of the occupied trees were dead, a highly significant difference ($\chi^2 = 8.04$, $P = 0.005$). In unoccupied trees herbivores attacked shoot meristems and leaves, and vines grew over the branches. Janzen concluded that the mortality rate of unoccupied trees was so great during the limited period of the experiments that the dependence of the trees on the ants was obligate. He also suggested that if all the trees of this species were abruptly removed from the area, then *P. ferruginea* would also crash to extinction, so dependent was it on the *Acacia* food supply and domatia.

A third study on the effects of the failure of ant service on plant populations was performed by Pudlo et al. (1980) on the ant-dispersed species *Sanguinaria canadensis*. Patterns of seed removal by ants, clone size, and sexual reproduction were analyzed for three populations, one in a frequently disturbed forest habitat subject to annual flooding and human use, a second in a forest that was cleared twenty-five years ago, and a third logged at the turn of the century and left undisturbed ever since. The three populations will be called disturbed, moderately disturbed, and undisturbed, respectively. The results are summarized in Table 18 and can be divided into two broad categories. First, the number of ant species taking seeds, the frequency of removals, and the dispersal distances were greater in the undisturbed habitat than in either of the disturbed habitats. Pudlo et al. (1980) argued that this was partially responsible for the second category of results, which concerned the demography of the populations: The number of *Sanguinaria* stems per square meter, the degree of aggregation of stems, the number of stems per clone, the mean number of seeds per capsule, and the proportion of stems with capsules were all lower in the undisturbed habitat.

There was little information on the history of the ant community at the moderately disturbed site, but a detailed list of the ant species occurring close to the most disturbed site was published in 1944. At that time about a dozen species known to disperse seeds were classified as either abundant or common. At the time of Pudlo's study only two species remained. The data show that the decimation of the ant community at the disturbed site seriously disrupted the ant–seed mutualism; fewer seeds were removed and they traveled very short distances. The importance of this was that although there was significantly more capsule and seed production at the disturbed site, seeds were not taken beyond the boundaries of the clones and were consequently buried at sites already occupied by adult plants. The probability of seedling establishment under these circumstances was low. Pudlo et al. (1980) further noted that Williams's (1975)

Table 18. *Differences in variables measured for three populations of* Sanguinaria canadensis *in habitats with different levels of disturbance*

Variables	Sites: Disturbed	Moderately disturbed	Undisturbed
Number of ant species removing seeds	3	1	5
Frequency of seed removals	0.07	0.56	0.64
Frequency of elaiosome-chewing with no removal	0.23	0.02	0.01
Mean seed dispersal distance	0.17m	0.89m	3.09m
Mean density of stems per m²	7.45	10.9	2.65
Dispersion, variance to mean ration (0 to ∞)	716.8	46.9	3.74
Dispersion, Lefkovitch measure (-1 to $+1$)	0.99	0.97	0.66
Mean number of stems per clone	10.5	12.9	2.8
Mean number of seeds per capsule	31.0	27.7	15.5
Proportion stems producing capsules	36%	–	25%

strawberry-coral model of sexual strategy applied to the *Sanguinaria* study, with one crucial exception. As the clones became very large, sexual reproduction and the production of the dispersible seeds became significantly more important. However, since the dispersal mechanism broke down, the impact of sexual reproduction was minimized. As Janzen (1974a) pointed out, species interactions may fail long before participating species become locally extinct. In the case of *Sanguinaria,* the population at the disturbed site may have been expanding by asexual propagation for decades, but if its seeds are not dispersed by mutualistic ants, it is most likely living on borrowed time.

In a final example, the possible extinction of not one but scores of ant-dispersed plant species has been forecast by Bond and Slingsby (1984) as a result of their studies of myrmecochory in the fynbos vegetation of Cape Province in South Africa. This is a fire-climax plant community, rich in ant-dispersed shrub species (Bond & Slingsby 1983). Many areas have been invaded by the introduced Argentine ant *Iridomyrmex humilis,* which has displaced indigenous ant species, including those that disperse seeds. Field experiments were carried out in which groups of fresh seeds of the myrmecochore *Mimetes cucullatus* (Proteaceae) were placed along transects in uninvaded and invaded habitats. In uninvaded habitats 80% of the seeds were removed within thirty minutes of the start of experiments and all of them were taken by the end of the day. *Anaplolepis custodiens* and *Pheidole capensis* were the principal ants involved and they relocated the seeds into their nests. By contrast, seeds in invaded habitats were removed at the much lower rate of 44% during the first day, and they were usually abandoned on or near the soil surface and not in nests.

The emergence of seedlings was studied in experimental and natural areas with and without Argentine ants. As seedling emergence is normally confined to postfire periods, observations were carried out after controlled burns. In experimental areas where 300 seeds had been "fed" into ant nests, invaded habitats yielded 1 out of 150 possible seedlings, whereas uninvaded habitats yielded 53 out of 150 possible seedlings. In areas with natural seedling emergence, the seedlings of uninvaded habitats were more numerous and larger than those in invaded habitats. In addition, most seedlings of invaded habitats emerged in those areas that, before the fires, were beneath the canopies of adult plants. As adults normally survive fires and sprout from lignotubers, these seedlings would be rapidly engulfed and shaded by masses of new shoots from established plants. In uninvaded habitats, however, the seedlings were widely scattered, suggesting that one advantage of ant dispersal that disappeared in the presence of *I. humilis* was a broad distribution of dispersal distances.

A second, and possibly more important, advantage of ant dispersal that was lost in invaded habitats was predator avoidance. Seed losses to small mammals, especially shrews, were significantly greater in invaded habitats, probably because *I. humilis* abandoned seeds exposed on the soil surface. Up to 72.5% of seeds were lost to predators at one site.

Bond and Slingsby showed that native seed-dispersing ant species relocated the majority of seeds to nests quickly before predators became a factor. The introduced ants, in displacing the mutualistic species, deprived *Mimetes* of ant services providing predator avoidance (and perhaps fire avoidance), and burial at microsites free of competition from adult plants. As a result there was little or no seedling recruitment in invaded habitats. This condition, if persistent, would lead to local extinctions of the plant species. The authors estimated that 170 more species in the Proteaceae alone may be in a similar predicament. This study is a clear demonstration of the importance of ants to the survival of a plant species and is indicative of the powerful selection that favors the evolution and maintenance of mutualistic traits.

All of the preceding examples involve major components of fitness such as seedling survivorship and seed production. They show that when mutualisms fail the consequences can be dire, at least for the plants. Selection for traits that maintain ant services can be intense, and this is a powerful *cohesive* force among the myriad variables affecting the evolution of mutualistic interactions.

A crucial behavioral factor is the tendency of many kinds of ant colonies to exhibit a density-dependent response to food (Wilson 1971; Carroll & Janzen 1973; Carroll & Risch 1983). Low numbers of plant reward items may not trigger a strong response, and ant services may be desultory and unsystematic. High concentrations of rewards such as pulses of extrafloral nectar during bud break and leaf expansion, or seed release, may foster a strong foraging response, often including recruitment or trail laying, which in turn increases the quality of ant services. The diversification of nutrients in the food rewards, and in cases such as *Acacia* the diversification of the rewards themselves, makes some sense in this context. Although the probability that plants bearing rewards benefit from the investment often appears to be small, or at best very variable, the rewards themselves are often potent and diverse. This may serve to maintain a minimal level of ant "interest" in the plants so that colonies keep in touch with the quantity and quality of plant rewards within reach. Thus, under ideal conditions, ant response to surges of rewards is immediate, and services are supplied when they are most needed.

Density-dependent responses may account for much of the variation in function and effect described throughout the previous chapters. The maintenance of ant–plant encounters at low frequencies may not provide easily measurable plant services for much of the time. However, they may enhance the probability that a plant benefits from a density response. In the case of plant protection mediated by the secretion of extrafloral nectar, damage from an outbreak or infestation of herbivores or seed predators may be significantly reduced by a density response on the part of the ants. The literature reviewed in Chapter 3 showed several examples in which ant protection was most effective against explosions of plant enemies that, in turn, stimulated major foraging episodes from surrounding ant colonies. In a similar vein, an abundant seed crop may produce a strong density response from seed-dispersing ants.

If some types of ant–plant mutualisms require resource concentration and density response, dramatic effects may be only occasional. This is likely to result in mixed findings from short-term field experiments and observations. However, a more important result is that the major selective advantage of a mutualism that provides ant services may reside in the response of ants to concentrations of resources such as prey or seeds. It is true that the secretion of extrafloral nectar can provide a constant low-level attrition of plant enemies, but the protective potential of ants may not be realized until the density of enemies reaches a critical level. Let us say that the outbreak of enemies threatens to remove the foliage from a plant or population of plants. Alternative methods of defense have clearly failed. A strong density response by ants may save the plant(s) from defoliation, and hence a decrement in competitive ability. A similar response to seed predators may save a seed crop. In the case of seed dispersal, a large seed crop resulting perhaps from an unusually favorable growing season is a major investment on the part of the plants. Field studies are complete enough to suggest that a strong density response by seed-dispersing ants will protect this investment by reducing losses to predators (which also may be density responsive), and by effectively dispersing the seeds to optimal microsites. In summary, ant–plant mutualisms of certain kinds, or under certain conditions, may have their greatest impact at unpredictable intervals. However, when this occurs the benefits may be enormous, such as the maintenance of competitive ability or the recruitment of seedlings. In both these examples the bottom line is the avoidance of local extinction.

Mutualistic benefits remain a question of probability. Elaiosomes do not guarantee that seeds will arrive at superior microsites for germination and establishment; they increase the probability of such an event. The presence of extrafloral nectaries does not guarantee ant protection; it increases the probability that protection will reach sufficient levels to maintain or increase fitness. Variation in every component of these interactions - genetic, demographic, ecological, and behavioral - generates variation in those probabilities.

When all components of a mutualism are present, the probability of benefits accruing to the plants and ants is at its greatest. Such an event may be infrequent and the contradictory results of field studies then emerge as symptoms of sampling error. The occurrence of a fully functioning mutualism in any given plant population is at present unpredictable either in space or time. However, greater sample sizes, in either dimension, might reduce somewhat the variation in results. Although it is reasonable to look for alternative hypotheses when contradictory results emerge, it is premature to reject the mutualistic hypothesis out of hand.

One extreme example is the physiological or "exploitationist" hypothesis for the function of extrafloral nectaries (Chapter 3). These organs may indeed function as some kind of physiological valve, but an emergent property of the mechanism is still protection by ants.

Variable and contradictory results may stem from many circumstances: The selection pressures that give rise to the adaptive response of ant-related traits – the plant enemies or the environmental stress – may be ameliorating or may have disappeared. This may have taken place as a result of demographic or geographical changes among the participating species. This argument might explain much of the variation in ant–plant mutualisms described in this and preceding chapters. It has also been suggested that some ant-related traits are relict and that at least some of the original adaptive partners are globally extinct. What we observe today is some derived or secondary interaction involving poorly adapted species. I can think of no test to discriminate between "primary" and "derived" mutualists when the "primary" ones are extinct!

Pathways to mutualistic interaction and coevolution

Mutualisms are widely held to evolve in response to stress, either physical or biological. Physical stress such as dessication or nutrient deficiency depresses the growth rates, survivorship, or fecundity of many species. Mutualisms arise following selection for traits that promote cooperation among species showing excessively low survivorship when "trying to make it on their own." Biological stress such as grazing or predation leads to the selection for traits that ameliorate the damage inflicted by interacting species. Thompson (1982) suggests that this is most likely to occur among species where individuals invariably experience damage during their lifetimes.

Ant–plant mutualisms may have often evolved as a result of some kind of stress. This has been discussed in previous chapters, but to summarize: Ant protection with the evolution of extrafloral nectaries, food bodies, and some types of domatia is a response to herbivores and seed predators (Table 19). Seed dispersal by ants may be a response to nutrient deficiency either inherent in the soil or the result of high nutrient demands by the plants. It is also a response to seed predation and fire. Finally, myrmecotrophy is also a response to nutrient deficiency. Mutualism has been a frequent response to nutrient deficiency in plants: Nitrogen-fixing symbionts in roots and the association of algae and fungi to produce lichens are common examples. Similarly, phytophagous insects that experience nutrient deficiency in their plant food frequently harbor gut symbionts.

Table 19. *Summary of plant resources, ant services, and selection pressures in ant–plant mutualisms*

	Services provided by ants, selection pressures involved		
Plant resources utilized by ants	Protection	Myrmecotrophy	Diaspore dispersal
Plant cavities	Herbivores, seed predators	Nutrient deficiency	Nutrient deficiency
Sap (from wounds, glands)	Herbivores, seed predators	NA	NA
Nectars	Herbivores, seed predators	?	?
Honeydew	Herbivores, seed predators	?	NA
Diaspores (seeds, fruits)	NA	NA	Nutrient deficiency, competition, fire, seed predators

Note: NA = probably not applicable.

The evolution of ant–plant mutualisms required that two crucial conditions be met: First, genetic variation existed for the mutualistic traits; and second, the traits produced resources (= rewards) not merely useful to ants, but whose utilization by ants increased the fitness of the plants that bore them. A corollary to this condition is that ants in incipient mutualisms might have benefited especially from freely accessible and abundant resources. Table 19 outlines the structure of most of the following discussion.

There is good evidence for the existence of genetic variation for mutualistic traits (see Chapters 3, 5, 6, 8). By contrast, the actual genetics of these traits is virtually unknown. Some, such as the nectariless trait of cotton, may be inherited by a relatively simple Mendelian mechanism (Meyer & Meyer 1961; Rhyne 1965). In fact, the presence or absence of a variety of epidermal structures involved in defense against herbivores is governed by just one or two genes (Gottleib 1984). These include spines, papillae, pubescence, trichomes, and extrafloral nectaries. Other traits on which the mutualism depends, such as perceptual behaviors, digestive abilities, and foraging, are likely to be the products of complex quantitative traits.

The evolution of mutualistic interactions has sometimes been viewed as a relatively simple series of reciprocating developments in response to the appearance of an advantageous gene in one of the participating species. Ecological conditions are likely to complicate this process. For example, a mutation resulting in the appearance of simple extrafloral nectaries might recruit the services of ants, which, in turn, increases the number of seeds bearing the nectary genes. By contrast, another likely scenario is that the nectar is consumed by organisms that in no way benefit the plant (or is not consumed at all), the extrafloral nectaries are a net drain on energy budgets, and the genes rapidly disappear. However, when ant-related plant traits are inherited by one or a few genes, and if the appearance of a trait in a natural population is an event of at least moderate probability, it is well to remember one of Haldane's (1932) major conclusions: A small selection coefficient or a small advantage to the plant can, given a sufficient population size, rapidly spread a gene throughout a species. One possible obstacle might be small population size, which would tend to lead to the loss of such genes. However, although population sizes in many plant species are small, predominant patterns of gene flow permit sufficient outbreeding for the spread of useful genes among populations (Levin & Kerster 1974). These patterns often appear to conform to the shifting-balance view of evolution (Wright 1932).

Turning now to the presumed quantitative mutualistic traits, especially in the ants, two factors might promote their evolution. First, as ants rarely specialize on plant rewards, the traits already mentioned such as foraging and perceptual behaviors and digestive characteristics will have been largely molded by the ants' general environment. In fact, it is possible that selection for plant reward traits has been to fit these molds. In other words, mutualistic responses by ants to plant rewards may not have required much adjustment of complex quantitative ant traits. This proviso may not be required, as quantitative traits may spread and be maintained under much the same conditions as Haldane outlined for simpler modes of inheritance (Lande 1976). Until more is known of the genetics of the very traits that determine the mutualistic interaction, we will remain with a very limited knowledge as to how they evolve.

The second condition required for an adaptive, mutualistic interaction is that the plant traits produce rewards that, when utilized by ants, create beneficial services for the plants. Most of the preceding chapters have included descriptions of how this functions in practice: For example, extrafloral nectar secretion does indeed increase the fitness of some plants, and elaiosomes generate a variety of ant-related benefits for the plants that bear them.

A crucial stage in the evolution of a mutualism is the first appearance of advantageous traits. Selection for these traits by potential partners may be weak (Howe 1984), and factors that promote their utilization could be important in increasing their frequency. Ant–plant mutualisms may have been favored by such factors, most nongenetic. First, as has already been pointed out in Chapter 3, plants may have been attractive to ant foragers because they harbored a variety of prey items in the form of plant chewers, borers, and suckers. The very presence of ants on plants would have promoted the use of the first plant rewards, which may have been weak or imperfect. This is an appropriate point at which to emphasize the importance of the presence of ants with preadapted behavioral repertoires. This has been illustrated by Koptur (1979), who showed that in California two nonnative species of *Vicia* (Papilionaceae) that bear extrafloral nectaries are readily protected by the introduced ant *I. humilis.* Although the ant has had no evolutionary experience with those species of *Vicia,* its entomophagous and nectarivorous behavior readily produced benefits to the plants. Second, as Scott (1980) suggested, tissues damaged by the activity of plant enemies may have released plant fluids attractive to ants. Some plant protection may have begun this way. Third, several authors have pointed out that honeydew is an abundant resource apparently underutilized by ants (Way 1963; Llewellyn 1972). Even in the presence of ants it can accumulate on surrounding foliage in sufficient quantities to form a sticky coat and glisten in the sun. There is no doubt that it is eventually metabolized by bacteria and fungi, but why ants do not harvest this seeming excess is unclear. To return to the main theme of this paragraph, it is possible that homopteran honeydew was an abundant reward on plants that increased ant presence and promoted homopteran-plant-ant interaction leading to some form of protection.

Thompson (1982) suggested that plant species with extrafloral nectaries have short-circuited this three-way interaction, diverting ant activity from the homopteran to their own tissues. Evidence presented in Chapters 1 and 2 suggests that both extrafloral nectaries and homopterans were present when angiosperms first evolved. An evolutionary "race" between plants with extrafloral nectaries and homopterans may be a partial explanation of the underutilization of honeydew by ants and does suggest that in competition with extrafloral nectaries, homopterans may have had to refine their ant-attracting capabilities.

How individual ant-related plant traits began is largely a mystery. With respect to domatia, Andrade and Carauta (1982) coined the term "metabiosis" to describe the process in which the development of plant structures fortuitously produces cavities suitable for nest sites. Risch et al. (1977) pointed out that in some plant genera new shoots are at first enclosed in protective sheaths. In some cases, such as *Piper,* the sheaths are not abscissed once the shoots mature and become attractive nest sites for some kinds of ants. However, the most complicated domatia are ontogenetically determined although ants modify them. Mutations that produce extrafloral nectaries are known, but we have little knowledge of what the first ones were like, or how ant behavior may have modified them. Elaiosomes are derived from several kinds of tissue, especially the funicle, which attaches the ovule to the ovary wall. A mutation resulting in this structure's being detached with the seed might result in the precursor of the elaiosome. The next stage would be the sequestering of ant rewards in the funicle. Many elaiosomes are vascularized and hence this stage may not have required any major structural reorganization.

Are ant–plant mutualisms coevolved? If we accept Schemske's (1983) definition of coevolution, "The joint selective effects on characters of interacting taxa, based on heritable variation in these characters," the answer is uncertain. There are examples of heritable variation in mutualistic traits as we have seen, and it is hard to imagine that an obligate relationship such as that between *Acacia* and *Pseudomyrmex* is anything but the result of powerful joint selection on traits producing particular plant rewards and particular ant behavior patterns. However, the majority of ant–plant mutualisms appear to be facultative, involving arrays of species, and this makes the question of coevolution cloudy.

The heart of the matter is the degree of *reciprocity* of selected traits. There is good evidence for variation in, and selection for, ant-related traits in plants. There is less evidence for analogous processes among the ant species involved in ant–plant mutualisms.

On the face of it the most obligate *Acacia–Pseudomyrmex* mutualisms appear to be cases of coevolution. Janzen (1966) listed sixteen specialized coevolved traits of *Acacia,* the most familiar of which are the enlarged foliar nectaries, the Beltian food bodies, and the soft-pithed thorns that form the domatia. He also listed specialized coevolved traits of *Pseudomyrmex* (Table 20). However, many of these traits are not confined to obligate acacia ants. The traits best explained by involvement with this particular mutualism are aggressiveness toward most animal intruders, the clipping of vegetation both on and around the host tree, and the high

Table 20. Pseudomyrmex *traits related to the ant–acacia coevolution (worker traits unless otherwise indicated)*

A. General features of *Pseudomyrmex* of importance to the interaction	B. Specialized features of obligate acacia ants (coevolved traits)
1. Fast and agile runners, not aggressive	1. Very fast and agile runners, aggressive
2. Good vision	2. Same as A.2
3. Independent foragers	4. Same as A.3
4. Smooth sting, barged sting sheath not inserted	4. Smooth sting, barbed sting sheath often inserted
5. Lick substrate, form buccal pellet	5. Same as A.5
6. Prey items retrieved entire	6. Same as A.6
7. Ignore living vegetation	7. Maul living vegetation contacting the swollen-thorn acacia
8. Workers without morphological castes	8. Same as A.8
9. Arboreal colony	9. Same as A.9
10. Highly mobile colony	10. Same as A.10
11. Larvae resistant to mortality by starvation	11. Same as A.11
12. One queen per colony	12. Sometimes more than one queen per colony
13. Colonies small	13. Colonies large
14. Diurnal activity outside nest	14. 24-hour activity outside nest
15. Few workers per unit plant surface	15. Many workers active on small plant surface area
16. Discontinuous food sources and unpredictable new nest site	16. Continuous food source and predictable new nest site
17. Founding queens forage far for food	17. Founding queens forage short distances for food
18. Not dependent on other species	18. Dependent on another species group

density of foragers per unit plant surface. Other traits such as polygyny in some obligate acacia ants may also be coevolved. Most of the others are common to ants that nest and forage in trees. Gleaning behavior (Table 20, item 5) is found in a variety of ant genera and even the use of Beltian bodies is most likely to be a simple response to a plant-borne attractant.

The subfamily Pseudomyrmecinae as a whole exhibits a great many traits that can be viewed as preadaptive in the context of mutualisms such as *Acacia–Pseudomyrmex* (Janzen 1966). Hence, another genus of the subfamily, *Pachysima,* interacts in a similar way with the tree *Barteria* (see Chapter 3). Once again, most of the traits exhibited by the ants do not appear to be specific to the mutualism with *Barteria.* However, a few are arguably coevolved: Janzen (1972) recorded how "there was literally a slow rain of ants from a heavily occupied *Barteria.*" This type of behavior may be explicable only in the context of defense against large mammals, which encounter the ant defenses at the periphery of the tree. This behavior has the appearance of a unique behavioral trait selected by a process of coevolution. Two other *Pachysima* traits may be of a similar nature: first, the tendency to search an enemy for a vulnerable spot before stinging, and second, the emission of a fetid odor, which may be a warning (Janzen 1972).

In both the *Acacia–Pseudomyrmex* and *Barteria–Pachysima* mutualisms there are many plant traits specifically related to the interaction with the ants. But the reverse is less obvious and coevolution appears to have been very lopsided. None of this is to say that more ant traits have not evolved in response to specific mutualistic plant species, but that the detection and analysis of such traits are extremely difficult. Even in the most obligate ant–plant mutualisms *reciprocal* selection is hard to demonstrate. Consider some ant-related plant traits such as food bodies, extrafloral nectar, and domatia. Selection for these traits by ants presumably leads to larger food bodies, more nectar, and better domiciles, which in turn result in superior ant services and ultimately increased plant fitness. But what about the ant traits? How does the plant increase the aggressiveness of ants toward animal intruders, invading vines, or nearby competing seedlings? The answer seems to be that the plant increases the attractiveness of its rewards to those ant species that already exhibit these behavioral characteristics. In other words, it can be argued that the plant is merely increasing the density of the most preadapted ant species on its surfaces. The plant is still "in the driver's seat." If this is the case then even obligate mutualisms are coevolved only in a highly restricted sense: most of the plant traits but few of the ant traits being specific to the mutualism.

The main reason for this is that the selection pressures have been asymmetrical, being applied mainly to the plants that have required the services of ants for crucial benefits such as protection or nutrients. Since many of the behavioral patterns required of the ants are common, and plant rewards are a variable component of ant diets, selection has been primarily on plant traits, especially rewards. The overwhelming majority of ant–plant mutualisms involve fortuitous groups or assemblages of ants responding to plant rewards. Ecological assemblages of ants respond to plant traits and may at times constitute the single most important selective factor in their evolution, but there are few signs of reciprocal selection and specialization among ants. Therefore, unlike the obligate minority, the majority of ant–plant mutualisms are diffuse and facultative. It can be argued that these mutualisms have arisen as a result of severe, directional, selective pressures on the plants in the presence of arrays of preadapted ant species. Although it is true that a few coevolved ant traits may be present in some obligatory mutualisms, there is little evidence for selective reciprocity among the majority of ant–plant mutualisms studied so far.

In summary, the majority of ant–plant mutualisms evolve in response to selection, especially stress selection, on plants. The participating ant species, which vary in time and space, respond to plant rewards on a facultative basis, and traits evolved in specific response to plant rewards are infrequent. Thus, when the definition of coevolution requires reciprocal selection involving heritable traits, coevolution is remarkably difficult to demonstrate in ant–plant mutualisms and appears to be the exception rather than the rule. In fact, the evidence suggests that it is directional selection that drives the evolution of the vast majority. This does not diminish the importance of their effects. The speed and energy with which ants harvest rewards such as elaiosomes suggest that they are crucial to the economy of ant colonies. This may be especially true when the colony is stressed, either by the external environment or by major internal demographic events such as reproduction. For the plants, we have repeatedly seen that ant services are potent forces that increase plant fitness. In many cases the failure of ant services leads to a variety of demographic failures, including poor seed set and poor seedling recruitment. Populations may contract, crash, or even become extinct as a result. Ant services, either on a continuous basis or as a density response, are crucial to a wide variety of plant species worldwide.

References

Abott, A. (1978). Nutrient dynamics of ants. In *Production Ecology of Ants and Termites,* ed. M. V. Brian, pp. 233–44. Cambridge: Cambridge University Press.

Addicott, J. F. (1979). A multispecies aphid-ant association: Density dependence and species-specific effects. *Canadian Journal of Zoology,* 57, 558–69.

Adlung, K. G. (1966). A critical evaluation of the European research on the use of red wood ants (*Formica rufa* group) for the protection of forests against harmful insects. *Zeitschrift für Angewandte Entomologie,* 57, 167–89.

Ahmad, S. (1983). *Herbivorous Insects.* New York: Academic Press.

Alexander, R. D. (1974). The evolution of social behavior. *Annual Review of Ecology and Systematics,* 5, 325–83.

Andersen, A. (1982). Seed removal by ants in the mallee of northwestern Victoria. In *Ant–Plant Interactions in Australia,* ed. R. C. Buckley, pp. 31–44. The Hague: Junk.

– (1983). Species diversity and temporal distribution of ants in the semi-arid mallee region of northwestern Victoria. *Australian Journal of Ecology,* 8, 127–37.

Andrade, J. C., de, & Carauta, J. P. P. (1982). The *Cecropia–Azteca* association: A case of mutualism? *Biotropica,* 14, 15.

Andrews, E. A. (1929). The mound-building ant *Formica exsectoides* F., associated with tree-hoppers. *Annals of the Entomological Society of America,* 22, 369–91.

Apostolov, L. G., & Likhovidov, V. E. (1973). Effect of red forest ants on oak leaf-roller pupae in the conditions of the south-western Ukraine. *Soviet Journal of Ecology,* 4, 178–9.

Applegate, R. D., Rogers, L. L., Casteel, D. A., & Novak, J. M. (1979). Germination of cow parsnip seeds from grizzly bear feces. *Journal of Mammalogy,* 60, 655.

Arber, E. A. N., & Parkin, J. (1907). On the origin of angiosperms. *Journal of the Linnean Society of London* (Botany), 38, 29–80.

Armstrong, J. A. (1979). Biotic pollination mechanisms in the Australian flora: A review. *New Zealand Journal of Botany,* 17, 467–8.

Atsatt, P. R. (1981a). Ant-dependent food plant selection by the mistletoe butterfly *Ogyris amaryllis* (Lycaenidae). *Oecologia,* 48, 60–3.

– (1981b). Lycaenid butterflies and ants: Selection for enemy-free space. *American Naturalist,* 118, 638–54.

Auclair, J. L. (1963). Aphid feeding and nutrition. *Annual Review of Entomology,* 8, 439–90.

Axelrod, D. I. (1970). Mesozoic paleogeography and early angiosperm history. *Botanical Reviews,* 36, 277–319.

Bacon, J. S. D., & Dickinson, B. (1957). The origin of melezitose: A biochemical relationship between the lime tree (*Tilia* spp.) and an aphid (*Eucallipterus tiliae* L.). *Biochemical Journal,* 66, 289–97.

Bailey, I. W. (1922a). The anatomy of certain plants from the Belgian Congo, with special reference to myrmecophytism. *Bulletin of the American Museum of Natural History,* 45, 585–621.

– (1922b). Notes on neotropical ant-plants. I. *Cecropia angulata. Botanical Gazette,* 74, 369–92.

– (1923). Notes on neotropical ant-plants. II. *Tachigalia paniculata* Aubl. *Botanical Gazette,* 75, 27–41.

– (1924). Notes on neotropical ant-plants. III. *Cordia nodosa* Lam. *Botanical Gazette,* 77, 32–49.

Baker, H. G., & Baker, I. (1975). Studies of nectar constitution and plant-pollinator coevolution. In *Coevolution of Animals and Plants,* ed. L. E. Gilbert & P. H. Raven, pp. 100–40. Austin: University of Texas Press.

– (1978). Ants and flowers. *Biotropica,* 10, 80.

– (1983). A brief historical review of the chemistry of floral nectar. In *The Biology of Nectaries,* ed. B. Bentley & T. Elias, pp. 126–52. New York: Columbia University Press.

Baker, H. G., Opler, P. A., & Baker, I. (1978). A comparison of the amino acid complements of floral and extrafloral nectars. *Botanical Gazette,* 139, 322–32.

Banks, C. J. (1962). Effects of the ant *Lasius niger* (L.) on insects preying on small populations of *Aphis fabae* Scop. on bean plants. *Annals of Applied Biology,* 50, 669–79.

Banks, C. J., & Macaulay, E. D. M. (1967). Effects of *Aphis fabae* Scop. and of its attendant ants and insect predators on yields of field beans. *Annals of Applied Biology,* 60, 445–53.

Barbier, M., & Delage, B. (1967). Le contenu des glandes pharyngiennes de la Fourmi *Messor capitatus* Latr. (Insecte Hyménoptère Formicidé). *Comptes Rendus de l'Academie de Sciences,* Paris, 264, 1520–2.

Baroni-Urbani, C., Josens, G., & Peakin, G. J. (1978). Empirical data and demographic parameters. In *Production Ecology of Ants and Termites,* ed. M. V. Brian, pp. 5–44. Cambridge: Cambridge University Press.

Bartlett, B. R. (1961). The influence of ants upon parasites, predators and scale insects. *Annals of the Entomological Society of America,* 54, 543–51.

Barton, A. M. (1983). The effects of heterogeneity on interactions between ants and an extrafloral nectary plant. *Bulletin of the Ecological Society of America,* 64, 118.

Baskin, S. I., & Bliss, C. A. (1969). Sugar occurring in the extrafloral exudates of the Orchidaceae. *Phytochemistry,* 8, 1139–45.

Bates, R. (1979). *Leporella fimbriata* and its ant pollinators. *Journal of the Native Orchid Society of South Australia,* 3, 9–10.

Baxendale, R. W. (1979). Plant-bearing coprolites from north American pennsylvanian coal balls. *Paleontology,* 22, 537–48.

Beadle, N. C. W. (1962). An alternative hypothesis to account for the generally low phosphate content of Australian soils. *Australian Journal of Agricultural Research,* 13, 434–42.

Beard, J. S. (1976). The evolution of Australian desert plants. In *The Evolution of Desert Biota,* ed. D. W. Goodall, pp. 51–63. Austin: University of Texas Press.

Beattie, A. J. (1971). Pollination mechanisms in *Viola. New Phytologist,* 70, 343–60.

– (1974). Floral evolution in *Viola. Annals of the Missouri Botanical Garden,* 61, 781–93.

– (1982). Ants and gene dispersal in flowering plants. In *Pollination and Evolution,* ed. J. A. Armstrong, J. M. Powell, & A. J. Richards, pp. 1–8. Sydney: Royal Botanic Gardens.

– (1983). Distribution of ant-dispersed plants. *Sonderbaende des Naturwissenschaftlichen Vereins in Hamburg,* 7, 249–70.

Beattie, A. J., & Culver, D. C. (1977). Effects of the mound nests of the ant *Formica obscuripes* on the surrounding vegetation. *American Midland Naturalist,* 97, 390–9.

– (1979). Neighborhood size in *Viola. Evolution,* 33, 1226–9.

– (1981). The guild of myrmecochores in the herbaceous flora of West Virginia forests. *Ecology,* 62, 107–15.

– (1982). Inhumation: How ants and other invertebrates help seeds. *Nature,* 297, 627.

– (1983). The nest chemistry of two seed-dispersing ant species. *Oecologia,* 56, 99–103.

Beattie, A. J., & Lyons, N. (1975). Seed dispersal in *Viola:* Adaptations and strategies. *American Journal of Botany,* 62, 714–22.

Beattie, A. J., Turnbull, C. L., Knox, R. B., & Williams, E. G. (1984). Ant inhibition of pollen function: A possible reason why ant pollination is rare. *American Journal of Botany,* 71, 421–6.

Beattie, A. J., Turnbull, C. L., Knox, R. B., & Hough, T. (in press). The vulnerability of pollen and fungal spores to ant secretions: Some evolutionary implications. *American Journal of Botany.*

Beckman, R. L., Jr., & Stucky, J. M. (1981). Extrafloral nectaries and plant guarding in *Ipomoea pandurata* (L.) (Convolvulaceae). *American Journal of Botany,* 68, 72–9.

Belt, T. (1874). *The Naturalist in Nicaragua.* London: Dent & Sons.

Bentley, B. L. (1977a). Extrafloral nectaries and protection by pugnacius bodyguards. *Annual Review of Ecology and Sysematics,* 8, 407–27.

– (1977b). The protective function of ants visiting the extrafloral nectaries of *Bixa orellana* (Bixaceae). *Journal of Ecology,* 65, 27–38.

– (1983). Nectaries in agriculture, with an emphasis on the tropics. In *The Biology of Nectaries,* ed. B. L. Bentley & T. Elias, pp. 204–22. New York: Columbia University Press.

Bequaert, J. (1922). Ants of the American Museum Congo Expedition. A contribution to the myrmecology of Africa. IV. Ants in their diverse relations to the plant world. *Bulletin of the American Museum of Natural History,* 45, 333–583.

Berg, R. Y. (1954). Development and dispersal of the seed *Pedicularis silvatica. Nytt Magasin (for) Botanikk,* 2, 1–59.

– (1958). Seed dispersal, morphology, and phylogeny of *Trillium. Skrifter Utgitt av Det Norske Videnskaps-Akademi, Oslo. 1.*

– (1966). Seed dispersal of *Dendromecon:* Its ecologic, evolutionary, and taxonomic significance. *American Journal of Botany,* 53, 61–73.

– (1975). Myrmecochorous plants in Australia and their dispersal by ants. *Australian Journal of Botany,* 23, 475–508.

Bernard, F. (1968). Les Fourmis d'Europe Occidentale et Septentrionale. *Faune de l'Europe et du Bassin Mediterraneen,* 3. Paris: Masson.

Bews, J. W. (1917). The plant succession in the thorn veld. *South African Journal of Science,* 14, 153–72.

Beyer, A. H. (1924). Life history of a new citrus aphid. *Florida Entomologist,* 8, 8–13.

Bhandal, I. S., & Malik, C. P. (1980). Total and polar lipid biosynthesis during *Crotalaria juncea* L. pollen tube growth: Effect of gibberellic acid, indoleacetic acid and (2-Chloroethyl-) phosphoric acid. *Journal of Experimental Botany,* 31, 931–5.

Bidwell, R. G. S. (1974). *Plant Physiology.* New York: Macmillan.

Bierzychudek, P. (1981). Pollinator limitation of plant reproductive effort. *American Naturalist,* 117, 838–40.

Blackwell, W. H., & Powell, M. J. (1981). A preliminary note on pollination in the Chenopodiaceae. *Annals of the Missouri Botanical Garden,* 68, 524–6.

Blom, P. E., & Clark, W. H. (1980). Observations of ants (Hymenoptera: Formicidae) visiting extrafloral nectaries of the barrel cactus, *Ferocactus gracilis* Gates (Cactaceae), in Baja California, Mexico. *Southwestern Naturalist,* 25, 181–96.

Boecklen, W. J. (1983). Experimental investigation of ant-plant mutualism in *Hibiscus aculeatus. Bulletin of the Ecological Society of America,* 64, 118.

Bond, W. J., & Slingsby, P. (1983). Seed dispersal by ants in shrublands of the Cape Province and its evolutionary implications. *South African Journal of Science,* 79, 231–3.

– (1984). Collapse of an ant-plant mutualism: The Argentine ant (*Iridomyrmex humilis*) and myrmecochorous Proteaceae. *Ecology,* 65, 1031–7.

Bormann, F. H., & Likens, G. E. (1979). *Pattern and Process in a Forested Ecosystem.* New York: Springer-Verlag.

Borror, D. J., Delong, D. M., & Triplehorn, C. A. (1976). *An Introduction to the Study of Insects.* Chicago: Holt, Rinehart and Winston.

Bowden, B. N. (1970). The sugars in the extrafloral nectar of *Andropogon gayanus* var. *bisquamulatus. Phytochemistry,* 9, 2315–18.

Bradley, G. A. (1972). Transplanting *Formica obscuripes* and *Dolichoderus tachenbergi* (Hymenoptera: Formicidae) colonies in jack pine stands of southeastern Manitoba. *Canadian Entomologist,* 104, 245–9.

Bradley, G. A., & Hinks, J. D. (1968). Ants, aphids, and jack pine in Manitoba. *Canadian Entomologist,* 100, 40–50.

Brantjes, N. B. M. (1981). Ant, bee and fly pollination in *Epipactis palustris* (L.) Crantz (Orchidaceae). *Acta Botanica Neerlandica,* 30, 59–68.

Bresinsky, A. (1963). Bau, Entwicklungsgeschichte und Inhaltsstoffe der Elaiosomen. *Bibliotheca Botanica,* 126, 1–54.

Brian, M. V. (1955). Food collection by Scottish ant community. *Journal of Animal Ecology,* 24, 336–51.

– (1956). Studies of caste differentiation in *Myrmica rubra* L. 4. Controlled larval nutrition. *Insectes Sociaux,* 3, 369–94.

– (1973). Caste control through worker attack in the ant *Myrmica. Insectes Sociaux,* 20, 87–107.

– (1977). *Ants.* London: Collins.

Brian, M. V., Hibble, J., & Kelly, A. F. (1966). The dispersion of ant species on a southern English heath. *Journal of Animal Ecology,* 35, 281–90.

Briese, D. (1974). Ecological studies on an ant community in a semiarid habitat, with emphasis on seed-harvesting species. Ph.D. thesis, Australian National University, Canberra.

Bristow, C. M. (1982). The structure of a temperate zone mutualism: Ants and Homoptera on New York ironweed (*Vernonia novoboracensis* L.). Ph.D. thesis, Princeton University.

Brown, H., & Martin, M. H. (1981). Pretreatment effects of cadmium on the root growth of *Holcus lanatus* L. *New Phytologist,* 89, 621–9.

Brown, W. L., Jr. (1954). Remarks on the internal phylogeny and subfamily classification of the family *Formicidae. Insectes Sociaux,* 1, 21–31.

– (1968). An hypothesis concerning the function of the metapleural glands in ants. *American Naturalist,* 102, 188–91.

– (1973). A comparison of the Hylean and Congo-West African rain forest ant faunas. In *Tropical Forest Ecosystems in Africa and South America: A Comparative Review,* ed. B. J. Meggers, E. S. Ayensu, & W. D. Duckworth, pp. 161–85. Washington, D.C.: Smithsonian Institution Press.

Brown, W. L., Jr., & Taylor, R. W. (1970). Superfamily Formicoidea. In *The Insects of Australia,* pp. 951–9. CSIRO: Melbourne University Press.

Buckley, R. (1983). Interaction between ants and membracid bugs decreases growth and seed set of host plant bearing extrafloral nectaries. *Oecologia,* 58, 132–6.

Bullock, S. H. (1974). Seed dispersal of *Dendromecon* by the seed predator *Pogonomyrmex. Madrono,* 22, 378–9.

Burger, W. C. (1981). Why are there so many kinds of flowering plants? *Bioscience,* 31, 572–81.

Burgett, D. M., & Young, R. G. (1974). Lipid storage by honey ant repletes. *Annals of the Entomological Society of America,* 67, 743–4.

Burns, D. P. (1964). Formicidae associated with tuliptree scale. *Annals of the Entomological Society of America,* 57, 137–9.

Burns, D. P., & Donley, D. E. (1969). Biology of the tuliptree scale, *Toumeyella liriodendri* (Homoptera: Coccidae). *Annals of the Entomological Society of America,* 63, 228–30.

Butler, G. D., Jr., Loper, G. M., McGregor, S. E., Webster, J. L., & Margolis, H. (1972). Amounts and kinds of sugars in the nectars of cotton. *Agronomy Journal,* 64, 364–8.

Carpenter, F. M. (1977). Geological history and the evolution of the insects. *Proceedings of the XV International Congress of Entomology,* pp. 63–70. College Park, Md.: The Entomological Society of America.

Carroll, C. R. (1979). A comparative study of two ant faunas: The stem-nesting ant communities of Liberia, West Africa and Costa Rica, Central America. *American Naturalist,* 113, 551–61.

Carroll, C. R., & Janzen, D. H. (1973). Ecology of foraging by ants. *Annual Review of Ecology and Systematics,* 4, 231–57.

Carroll, C. R., & Risch, S. J. (1983). Tropical annual cropping systems: Ant colonies. *Environmental Management,* 7, 65–72.

Cavill, G. W. K., Davies, N. W., & McDonald, F. J. (1980). Characterization of aggregation factors and associated compounds from the Argentine ant, *Iridomyrmex humilis. Journal of Chemical Ecology,* 6, 371–84.

Cavill, G. W. K., & Robertson, P. L. (1965). Ant venoms, attractants and repellents. *Science,* 149, 1337–45.

Cavill, G. W. K., Robertson, P. L., & Whitfield, F. B. (1964). Venom and venom apparatus of the bull ant, *Myrmecia gulosa* (Fabr.). *Science,* 146, 79–80.

Cherrett, J. M. (1972). Chemical aspects of plant attack by leaf-cutting ants. In *Phytochemical Ecology,* ed. J. B. Harborne, pp. 13–24. London: Academic Press.

Chippendale, G. M. (1972). Insect metabolism and dietary sterols and essential fatty acids. In *Insect and Mite Nutrition,* ed. J. G. Rodriguez, pp. 423–35. Amsterdam: North-Holland.

Clark, E. W., & Lukefahr, M. J. (1956). A partial analysis of cotton extrafloral nectar and its approximation as a nutritional medium for adult pink bollworms. *Journal of Economic Entomology,* 49, 875–6.

Clayton, R. B. (1964). The utilization of sterols by insects. *Journal of Lipid Research,* 5, 3–19.

Cole, A. C. (1940). A guide to the ants of the Great Smoky Mountains National Park, Tennessee. *American Midland Naturalist,* 24, 1–88.

Coley, P. D. (1980). Effects of leaf age and plant life history patterns on herbivory. *Nature,* 284, 545–6.

– (1983). Herbivory and defensive characteristics of tree species in a lowland tropical forest. *Ecological Monographs,* 53, 209–33.

– (1983). Intraspecific variation in herbivory on two tropical tree species. *Ecology,* 64, 426–33.

Colombel, P. (1970). Recherches sur la biologie, et l'éthologie d'*Odontomachus haematodes* L. (Hym. Formicoidea, Ponerinae). Étude des populations dans leur milieu naturel. *Insectes Sociaux,* 17, 183–98.

Cook, R. E. (1979). Patterns of juvenile mortality and recruitment in plants. In *Topics in Plant Population Biology,* ed. O. T. Solbrig, pp. 201–31. New York: Columbia University Press.

– (1980). Germination and size-dependent mortality in *Viola blanda. Oecologia,* 47, 115–17.

Crepet, W. L. (1979). Insect pollination: A paleontological perspective. *Bioscience,* 29, 102–8.
Crowson, R. A., Rolfe, W. D. I., Smart, J., Waterson, C. D., Willey, E. C., & Wootton, R. J. (1967). Insecta. In *The Fossil Record,* ed. W. B. Harland et al., pp. 508–28. Geological Society of London. London: Burlington House.
Culver, D. C., & Beattie, A. J. (1978). Myrmecochory in *Viola:* Dynamics of seed-ant interactions in some West Virginia species. *Journal of Ecology,* 66, 53–72.
- (1980). The fate of *Viola* seeds dispersed by ants. *American Journal of Botany,* 67, 710–14.
- (1983). Effects of ant mounds on soil chemistry and vegetation patterns in a Colorado montane meadow. *Ecology,* 64, 485–92.
Czerwinski, Z., Jakubczyk, H., & Petal, J. (1969). The influence of ants of the genus *Myrmica* on the physico-chemical and microbiological properties of soil within the compass of ant-hills in the Strzeleckie Meadows. *Polish Journal of Soil Science,* 3, 51–8.
- (1971). Influence of ant hills on meadow soils. *Pedobiolgia,* 11, 277–85.
Dadd, R. H. (1973). Insect nutrition: Current developments and metabolic implications. *Annual Review of Entomology,* 18, 381–420.
Dadd, R. H., & Krieger, D. L. (1967). Continuous rearing of aphids of the *Aphis fabae* complex on sterile synthetic diet. *Journal of Economic Entomology,* 60, 1512–14.
Dahl, E., & Hadac, E. (1940). Maur som blomsterbestvere. *Nytt Magasin Naturvidensk,* 81, 46.
Davidson, D. W., & Morton, S. R. (1981a). Competition for dispersal in ant-dispersed plants. *Science,* 213, 1259–61.
- (1981b). Myrmecochory in chenopodiaceous plants of the Australian arid zone. *Oecologia,* 50, 357–66.
Davison, E. A. (1982). Seed utilization by harvester ants. In *Ant–Plant Interactions in Australia,* ed. R. C. Buckley, pp. 1–6. The Hague: Junk.
DeBach, P., & Huffaker, C. B. (1971). Experimental techniques for evaluation of the effectiveness of natural enemies. In *Biological Control,* ed. C. B. Huffaker, pp. 113–40. New York: Plenum Press.
Delevoryas, T. (1962). *Morphology and Evolution of Fossil Plants.* New York: Holt, Rinehart and Winston.
Delpino, F. (1886). Funzione mirmecofila nel regno vegetale. *Memorie della R. Academia delle Scienze dell instituto di Bologna, Serie Quatro Tomo,* 8, 215–323.
Dethier, V. G. (1947). *Chemical Insect Attractants and Repellents.* Philadelphia: Blakiston.
Detling, J. K., & Dyer, M. I. (1981). Evidence for potential plant growth regulators in grasshoppers. *Ecology,* 62, 485–8.
Deuth, D. (1977). The function of extrafloral nectaries in *Aphelandra deppeana* Sch. and Cham., (Acanthaceae). *Brenesia,* 10, 135–45.
Dickison, W. C. (1975). The bases of angiosperm phylogeny: Vegetative anatomy. *Annals of the Missouri Botanical Garden,* 62, 590–620.

Dixon, A. F. G. (1971). The role of aphids in wood formation. I. The effect of the Sycamore aphid *Drepanosiphum platanoides* (Schr.) (Aphididae), on the growth of Sycamore, (*Acer pseudoplantanus L.*). *Journal of Applied Ecology,* 8, 165–79.

Downer, R. G. H., & Matthews, J. R. (1976). Patterns of lipid distribution and utilization in insects. *American Zoologist,* 16, 733–45.

Downhower, J. F. (1975). The distribution of ants on *Cecropia* leaves. *Biotropica,* 7, 59–62.

Doyle, J. A. (1977). Patterns of evolution in early angiosperms. In *Patterns of Evolution as Illustrated by the Fossil Record,* ed. A. Hallam, pp. 501–25. New York: Elsevier.

– (1978). Origin of angiosperms. *Annual Review of Ecology and Systematics,* 9, 365–92.

Doyle, J. A., & Hickey, L. J. (1974). Pollen and leaves from the mid-Cretaceous Potomac groups and their bearing on early angiosperm evolution. In *The Origin and Early Evolution of Angiosperms,* ed. C. B. Beck, pp. 139–206. New York: Columbia University Press.

Drake, W. E. (1981). Ant-seed interaction in dry sclerophyll forest of north Stradbroke island, Queensland. *Australian Journal of Botany,* 29, 293–309.

Duviard, D. (1969). Importance de *Veronia quineensis* Benth. dans l'alimentation de quelques fourmis de savane. *Insectes Sociaux,* 16, 115–34.

Duviard, D., & Segeren, P. (1974). La colonisation d'un myrmécophyte, le parasolier par *Crematogaster* spp. (Myrmicinae) en Cote-D'Ivoire forestière. *Insectes Sociaux,* 21, 191–212.

Dyer, M. I., & Bokhari, V. G. (1976). Plant-animal interactions: Studies of the effects of grasshopper grazing on blue grama grass. *Ecology,* 57, 762–72.

Dymes, T. A. (1916). The seed-mass and dispersal of *Helleborus foetidus. Journal of the Linnean Society of London (Botany),* 43, 433–55.

Eastop, V. F. (1973). Deductions from the present day host plants of aphids and related insects. In *Insect–Plant Relationships,* ed. H. F. van Emden, pp. 158–78. Oxford: Blackwell Scientific Publications.

Edinger, B. B. (1983). A multi-species ant-aphid association: Ant behavior differences correlate with tending benefit. *Bulletin of the Ecological Society of America,* 64, 67.

Ehrenfeld, J. (1978). Pollination of three species of *Euphorbia* subgenus *Chamesyce,* with special reference to bees. *American Midland Naturalist,* 101, 87–98.

Ehrlich, P. R., & Raven, P. J. (1964). Butterflies and plants: A study in coevolution. *Evolution,* 18, 581–608.

Eisner, T. (1957). A comparative morphological study of the proventriculus of ants (Hymenoptera: Formicidae). *Bulletin of the Museum of Comparative Zoology, Harvard University,* 116, 439–90.

Eisner, T., & Happ, G. M. (1962). The infrabuccal pocket of a formicine ant: A social filtration device. *Psyche,* 69, 107–16.

Eisner, T., Jutro, P., Aneshansley, D. J., & Niedhauk, R. (1972). Defense against ants in a caterpillar that feeds on ant-guarded scale insects. *Annals of the Entomological Society of America,* 65, 987–8.

Elias, T. S. (1983). Extrafloral nectaries: Their structure and distribution. In *The Biology of Nectaries,* ed. B. L. Bentley and T. S. Elias, pp. 174–203. New York: Columbia University Press.

Elias, T. S., Rozich, W. R., & Newcombe, L. (1975). The foliar and floral nectaries of *Turnera ulmifolia* L. *American Journal of Botany,* 62, 570–6.

Elmes, G. W., & Wardlaw, J. C. (1982). A population study of the ants *Myrmica sabuleti* and *M. scarbrinodis* living at two sites in the south of England. 2. Effect of above-nest vegetation. *Journal of Animal Ecology,* 51, 665–80.

Elton, C. S. (1932). Territory among wood ants (*Formica rufa L.*) at Picket Hill. *Journal of Animal Ecology,* 1, 69–76.

El-Ziady, S. (1960). Further effects of *Lasius niger* L. on *Aphis fabae* Scopoli. *Proceedings of the Royal Entomological Society of London (A),* 35, 30–8.

Ernst, N. H. O. (1980). Population biology of *Allium ursinum* in northern Germany. *Journal of Ecology,* 67, 347–60.

Evans, H. C. (1974). Natural control of arthropods, with special reference to ants (Formicidae), by fungi in the tropical high forest of Ghana. *Journal of Applied Ecology,* 11, 37–49.

Evans, H. E., & Eberhard, M. J. W. (1970). *The Wasps.* Ann Arbor: University of Michigan Press.

Ewart, W. H., & Metcalfe, R. L. (1956). Preliminary studies of sugars and amino acids in the honeydews of five species of coccids feeding on citrus in California. *Annals of the Entomological Society of America,* 49, 441–7.

Faegri, K., & van der Pijl, L. (1966). *The Principles of Pollination Ecology.* Toronto: Pergamon Press.

Feeny, P. (1976). Plant apparency and chemical defenses. *Recent Advances in Phytochemistry,* 10, 1–42.

Feinsinger, P., & Swarm, L. A. (1978). How common are ant-repellent nectars? *Biotropica,* 10, 238–9.

Finnegan, R. J. (1975). Introduction of a predacious red wood ant, *Formica lugubris* (Hymenoptera: Formicidae), from Italy to Eastern Canada. *Canadian Entomologist,* 107, 1271–4.

– (1977). Establishment of a predacious red wood ant, *Formica obscuripes* (Hymenoptera: Formicidae), from Manitoba to Eastern Canada. *Canadian Entomologist,* 109, 1145–8.

Flanders, S. E. (1957). The complete interdependence of an ant and a coccid. *Ecology,* 38, 535–6.

Freeland, J. (1958). Biological and social patterns in the Australian bulldog ants of the genus *Myrmecia. Australian Journal of Zoology,* 6, 1–18.

Frey-Wyssling, A. (1955). The phloem supply to the nectaries. *Acta Botanica Neerlandica,* 4, 358–69.

Friend, W. G. (1958). Nutritional requirements of phytophagous insects. *Annual Review of Entomology,* 3, 57–74.

Fritz, R. S. (1982). An ant-treehopper mutualism: Effects of *Formica subsericea* on the survival of *Vanduzea arguata. Ecological Entomology,* 7, 267–76.

Fritz, R. S., & Morse, D. H. (1981). Nectar parasitism of *Asclepias syriaca* by ants: Effect on nectar levels, pollinia insertion, pollinaria removal and pod

production. *Oecologia,* 50, 316–19.
Galen, C. (1983). The effects of nectar thieving ants on seedset in floral scent morphs of *Polemonium viscosum. Oikos,* 41, 245–9.
Gentry, J. G., & Stiritz, K. L. (1972). The role of the Florida harvester ant in old field mineral nutrient relationships. *Environmental Entomology,* 1, 39–41.
Gilbert, L. E. (1975). Ecological consequences of a coevolved mutualism between butterflies and plants. In *Coevolution of Animals and Plants,* ed. L. E. Gilbert & P. H. Raven, pp. 210–40. Austin: University of Texas Press.
– (1980). Food web organization and the conservation of neotropical diversity. In *Conservation Biology,* ed. M. E. Soule & B. A. Wilcox, pp. 11–33. Sunderland, Mass.: Sinauer.
Gilbert, L. E., & Smiley, J. T. (1978). Determinants of local diversity in phytophagous insects: Host specialists in tropical environments. *Symposium of the Royal Entomological Society London,* 9, 89–104.
Gilbert, L. I. (1967). Lipid metabolism and function in insects. *Advances in Insect Physiology,* 4, 69–211.
Gilbert, L. I., & Chino, H. (1974). Transport of lipids in insects. *Journal of Lipid Research,* 15, 439–56.
Gislen, T. (1949). Problems concerning the occurrence of *Melampyrum arvense* in Sweden. *Oikos,* 1, 115–31.
Givnish, T. J., Terborgh, J. W., & Waller, D. M. (in press). Plant form, temporal community structure, and species richness in forest herbs of the Virginia piedmont. *Ecological Monographs.*
Glancey, B. M., Stringer, C. E., Jr., Craig, C. H., Bishop, P. M., & Martin, B. B. (1973). Evidence of a replete caste in the fire ant *Solenopsis invicta. Annals of the Entomological Society of America,* 66, 233–4.
Glancey, B. M., Vander Meer, R. K., Glover, A., Lofgren, C. S., & Vinson, S. B. (1979). Filtration of microparticles from liquids ingested by the red imported fire ant *Solenopsis invicta* Buren. *Insectes Sociaux,* 28, 395–401.
Gomez, L. D. (1974). Biology of the potato fern. *Brenesia,* 4, 37–61.
Gosswald, K. (1951). Die rote Waldameise in Dienste der Waldhygiene: *Forstwirtschaftliche Bedeutung, Nutzung, Lebensweise, Zucht, Vermehrung und Schutz.* Luneburg: Melta Kinau.
Gosswald, K., & Horstmann, K. (1966). Untersuchungen über den Einfluss der kleinen roten Waldameise (*Formica polyctena* Foerster) auf den Massenwechsel des grünen Eichenwicklers (*Tortrix viridana* L.). *Waldhygiene,* 6, 230–55.
Gosswald, K., & Kloft, W. (1956). Untersuchungen über die Verteilung von radioaktiv markiertem Futter im Volk der kleinen roten Waldameise (*Formica rufopratensis minor*). *Waldhygiene,* 1, 200–2.
Gottlieb, L. D. (1984). Genetics and morphological evolution in plants. *American Naturalist,* 123, 681–709.
Gotwald, W. H. (1969). Comparative morphological studies of ants, with particular reference to the mouthparts (Hymenoptera: Formicidae). *Memoirs of Cornell University Agricultural Experimental Station,* Ithaca, N.Y., 408, 1–150.

Gray, B. (1971). Notes on the field behavior of two ant species *Myrmecia desertorum* Wheeler and *Myrmecia dispar* (Clark) (Hymenoptera: Formicidae). *Insectes Sociaux,* 18, 81–94.

Gray, R. A. (1952). Composition of honeydew excreted by pineapple mealybugs. *Science,* 115, 129–33.

Greenslade, P. J. M. (1971). Interspecific competition and frequency changes among ants in Solomon Islands coconut plantations. *Journal of Applied Ecology,* 8, 323–49.

Greenslade, P. J. M., & Halliday, R. B. (1983). Colony dispersion and relationships of meat ants *Iridomyrmex purpureus* and allies in an arid locality in South Australia. *Insectes Sociaux,* 30, 82–99.

Grimalsky, V. I. (1960). On the role of the ant *Formica rufa* in forest biocoenoses in the eastern Polessye of the Ukraine. *Zoologicheskii Zhurnal,* 39, 398.

Grime, J. P. (1979). *Plant Strategies and Vegetation Processes.* New York: Wiley.

Groff, G. W., & Howard, C. W. (1925). The cultured citrus of South China. *Lingnan Science Journal,* 2, 108–14.

Gutschick, V. P. (1981). Evolved strategies in nitrogen acquisition by plants. *American Naturalist,* 118, 607–37.

Hagerup, O. (1943). Myre-bestovning. *Botanisk Tidsskrift,* 46, 116–23.

Haines, B. (1975). Impact of leaf-cutting ants on vegetation development at Barro Colorado Island. In *Tropical Ecosystems: Trends in Terrestrial and Aquatic Research,* ed. F. D. Golley & E. Medina, pp. 99–111. New York: Springer-Verlag.

Haldane, J. B. S. (1932). *The Causes of Evolution.* Ithaca: Cornell University Press.

Hamilton, W. D. (1972). Altruism and related phenomenon, mainly in the social insects. *Annual Review of Ecology and Systematics,* 3, 193–232.

Handel, S. N. (1976). Dispersal ecology of *Carex pedunculata* (Cyperaceae), a new North American myrmecochore. *American Journal of Botany,* 63, 1071–9.

– (1978). The competitive relationship of three woodland sedges, and its bearing on the evolution of ant dispersal of *Carex pedunculata. Evolution,* 32, 151–63.

Handel, S. N., Fisch, S. B., & Schatz, G. E. (1981). Ants disperse a majority of herbs in a mesic forest community in New York State. *Bulletin of the Torrey Botanical Club,* 108, 430–7.

Harborne, J. B. (1978). *Biochemical Aspects of Plant and Animal Coevolution.* London: Academic Press.

Harper, J. L. (1977). *Population Biology of Plants.* London: Academic Press.

Harris, P. (1973). Insects in the population dynamic of plants. In *Insect-Plant Relationships,* ed. H. F. van Emden, pp. 201–10. Oxford: Blackwell Scientific Publications.

Harris, T. M. (1964). *The Yorkshire Jurassic Flora,* 2. *Caytoniales, Cycadales & Pteridosperms.* London: British Museum (Natural History).

Haskins, C. P., & Haskins, E. F. (1950). Notes on the biology and social behavior of the archaic Ponerine ants of the genera *Myrmecia* and *Promyrmecia. Annals of the Entomological Society of America,* 43, 461–91.

Haskins, C. P., & Wheldon, R. M. (1954). Notes on the exchange of ingluvial food in the genus *Myrmecia*. *Insectes Sociaux*, 1, 33–7.

Headley, A. E. (1952). Colonies of ants in locust woods. *Annals of the Entomological Society of America*, 45, 435–42.

Heinrich, B. (1978). The economics of insect sociality. In *Behavioral Ecology: An Evolutionary Approach*, ed. J. R. Krebs & N. B. Davies, pp. 97–128. Sunderland, Mass.: Sinauer.

– (1979). *Bumblebee Economics*. Cambridge: Harvard University Press.

Heithaus, E. R. (1981). Seed predation by rodents on three ant-dispersed plants. *Ecology*, 62, 136–45.

– (1983). Fates of frugivore-dispersed seeds. *Bulletin of the Ecological Society of America*, 64, 53.

Hendry, L. B., Wichmann, J. K., Hindenlang, D. M., Mumma, R. O., & Anderson, M. E. (1975). Evidence for the origin of insect sex pheromones: Presence in food plants. *Science*, 188, 59–63.

Hendry, L. B., Wichmann, J. K., Hindenlang, D. M., Weaver, K. M., & Korzeniowski, S. H. (1976). Plants – the origin of kairomones utilized by parasitoids of phytophagous insects? *Journal of Chemical Ecology*, 2, 271–83.

Herbers, J. M. (1979). Caste-biased polyethism in a mound-building ant species. *American Midland Naturalist*, 101, 69–75.

Hetschko, A. (1908). Ueber den Insektenbesuch bei einigen *Vicia* – Arten mit extrafloren Nektarien. *Wiener Entomologische Zeitung*, 27, 299–305.

Hickey, L. J., & Wolfe, J. A. (1975). The bases of angiosperm phylogeny: Vegetative morphology. *Annals of the Missouri Botanical Garden*, 62, 538–89.

Hickman, J. C. (1974). Pollination by ants: A low-energy system. *Science*, 184, 1290–2.

Hinton, H. E. (1951). Myrmecophilous Lycaenidae and other Lepidoptera – A summary. *Transactions of the South London Entomological and Natural History Society*, 1949–50, 111–75.

– (1977). Subsocial behaviour and biology of some Mexican membracid bugs. *Ecological Entomology*, 2, 61–79.

Hocking, B. (1970). Insect associations with the swollen thorn acacias. *Transactions of the Royal Entomological Society of London*, 122, 211–55.

– (1975). Ant-plant mutualism: Evolution and energy. In *Coevolution of Animals and Plants*, ed. L. E. Gilbert & R. H. Raven, pp. 78–90. Austin: University of Texas Press.

Holldobler, B. (1980). Canopy orientation: A new kind of orientation in ants. *Science*, 210, 86–8.

Holldobler, B., & Lumsden, C. J. (1980). Territorial strategies in ants. *Science*, 210, 732–9.

Holldobler, B., & Wilson, E. O. (1977). Weaver ants: Social establishment and maintenance of territory. *Science*, 195, 900–2.

Holttum, R. E. (1954a). *Plant Life in Malaya*. London: Longmans, Green.

– (1954b). *A Revised Flora of Malaya*. Vol. 2, *Ferns of Malaya*. Singapore: Government Printing Office.

Hori, K. (1976). Plant growth-regulating factor in the salivary gland of several Heteropterous insects. *Comparative Biochemistry and Physiology*, 53B, 435–8.

Horn, H. S. (1971). *The Adaptive Geometry of Trees.* Princeton, N.J.: Princeton University Press.

Horstmann, K. (1972). Investigations on the food consumption of red wood ants (*Formica polyctena* Foerster) in an oak forest. 2. Effect of season and supply of food. *Oecologia,* 8, 371–90.

– (1974). Untersuchungen über den Nahrungserwerb der Waldameisen (*Formica polyctena* Foerster) im Eichenwald. *Oecologia,* 15, 187–204.

Horvitz, C. C. (1981). Analysis of how ant behaviors affect germination in a tropical myrmecochore *Calathea microcephala* (P. & E.) Koernicke (Marantaceae): Microsite selection and aril removal by neotropical ants, *Odontomachus, Pachycondyla,* and *Solenopsis* (Formicidae). *Oecologia,* 51, 47–52.

Horvitz, C. C., & Beattie, A. J. (1980). Ant dispersal of *Calathea* seeds by carnivorous ponerines in a tropical rain forest. *American Journal of Botany,* 67, 321–6.

Horvitz, C. C., & Schemske, D. W. (1984). Effects of ant-mutualists and an ant-sequestering herbivore on seed production of a tropical herb *Calathea ovandensis* (Marantaceae). *Ecology,* 65, 1369–78.

Houk, E. J., & Griffiths, G. W. (1980). Intracellular symbiotes of the Homoptera. *Annual Review of Entomology,* 25, 161–87.

Howard, R. W., & Blomquist, G. J. (1982). Chemical ecology and biochemistry of insect hydrocarbons. *Annual Review of Entomology,* 27, 149–72.

Howe, H. F. (1984). Constraints on the evolution of mutualisms. *American Naturalist,* 123, 764–77.

Hsiao, K.-J. (1980). The use of biological agents for the control of the pine defoliator, *Dendrolimus punctatus* (Lepidoptera, Lasiocampidae), in China. *Protection Ecology,* 2, 297–303.

Huffaker, C. B. (1971). *Biological Control.* New York: Plenum Press.

Hutchinson, G. L., Millington, R. J., & Peters, D. B. (1972). Atmospheric ammonia: Absorption by plant leaves. *Science,* 175, 771–2.

Huxley, C. R. (1978). The ant-plants *Myrmecodia* and *Hydnophytum* (Rubiaceae) and the relationships between their morphology, ant occupants, physiology and ecology. *New Phytologist,* 80, 231–68.

– (1980). Symbiosis between ants and epiphytes. *Biological Reviews,* 55, 321–40.

Inouye, D. W. (1983). The ecology of nectar robbing. In *The Biology of Nectaries,* ed. B. Bentley and T. Elias, pp. 153–73. New York: Columbia University Press.

Inouye, D. W., & Taylor, O. R. (1979). A temporate region plant-ant-seed predator system: Consequences of extrafloral nectar secretion by *Helianthella quinquinervis. Ecology,* 60, 1–7.

Inozemtsev, A. A. (1974). Dynamics of the food relationships of common red ants and their role in regulating the abundance of noxious invertebrates in oak forests of the Tula district. *Soviet Journal of Ecology,* 5, 252–62.

Iwanami, Y., & Iwadare, T. (1978). Inhibiting effects of myrmicacin on pollen growth and pollen tube mitosis. *Botanical Gazette,* 139, 42–5.

– (1979). Myrmic acids: A group of new inhibitors analogous to myrmicacin. *Botanical Gazette,* 140, 1–4.

Iwanami, Y., Nakamura, S., Miki-Hirosige, H., & Iwadare, T. (1981). Effects of myrmicacin (β-hydroxydecanoic acid) on protoplasmic movement and ultrastructure of *Camellia japonica* pollen. *Protoplasma,* 104, 341–5.

Iwanami, Y., Okada, I., Iwamatsu, M., & Iwadare, T. (1979). Inhibitory effects of royal jelly acid, myrmicacin, and their analgous compounds on pollen germination, pollen tube elongation, and pollen tube mitosis. *Cell Structure and Function,* 4, 135–43.

Jacot-Guillarmod, C. (1951). A south African leguminous plant attractive to Hymenoptera. *Entomologists Monthly Magazine,* 87, 235–6.

James, W. O. (1950). Alkaloids in the plant. In *The Alkaloids: Chemistry and Physiology,* ed. R. H. F. Manske & H. L. Holmes, vol. 1, pp. 116–90. New York: Academic Press.

Janos, D. P. (1980). Vesicular-arbuscular mycorrhizae affect lowland tropical rain forest plant growth. *Ecology,* 61, 151–62.

Janzen, D. H. (1965). The interaction of the Bull's-horn acacia (*Acacia cornigera* L.) with one of its ant inhabitants (*Pseudomyrmex fulvescens* Emery) in eastern Mexico. Ph.D. thesis, University of California, Berkeley.

– (1966). Coevolution of mutualism between ants and acacias in central America. *Evolution,* 20, 249–75.

– (1967). Fire, vegetation structure and the ant-*Acacia* interaction in central America. *Ecology,* 48, 26–35.

– (1969a). Allelopathy by myrmecophytes: The ant *Azteca* as an allelopathic agent of *Cecropia. Ecology,* 50, 147–53.

– (1969b). Seed eaters versus seed size, number, toxicity and dispersal. *Evolution,* 23, 1–27.

– (1972). Protection of *Barteria* (Passifloraceae) by *Pachysima* ants (Pseudomyrmecinae) in a Nigerian rain forest. *Ecology,* 53, 885–92.

– (1973a). Dissolution of mutualism between *Cecropia* and its *Azteca* ants. *Biotropica,* 5, 15–28.

– (1973b). Evolution of polygynous obligate acacia-ants in Western Mexico. *Journal of Animal Ecology,* 42, 727–50.

– (1974a). The deflowering of Central America. *Natural History,* 83, 48–53.

– (1974b). Swollen-thorn acacias of Central America. *Smithsonian Contributions to Botany,* 13, 1–131.

– (1974c). Epiphytic myrmecophytes in Sarawak: Mutualism through the feeding of plants by ants. *Biotropica,* 6, 237–59.

– (1977). Why don't ants visit flowers? *Biotropica,* 9, 252.

– (1979). New horizons in the biology of plant defenses. In *Herbivores: Their Interaction with Secondary Plant Metabolites,* ed. G. A. Rosenthal & D. H. Janzen, pp. 331–50. New York: Academic Press.

– (1981). *Enterolobium cyclocarpum* seed passage rate and survival in horses, Costa Rican Pleistocene seed dispersal agents. *Ecology,* 62, 593–601.

Jeffrey, D. C., Arditti, J., & Koopowitz, H. (1970). Sugar content in floral and extrafloral exudates of orchids: Pollination, myrmecology and chemotaxonomy implication. *New Phytologist,* 69, 187–95.

Jones, C. R. (1929). Ants and their relation to aphids. *Colorado Experiment Station Bulletin* (Fort Collins), 341, 5–96.

Kawano, S. (1975). The productive and reproductive biology of flowering plants. 2. The concept of life history strategy in plants. *Journal of the College of Liberal Arts. Toyama University (Natural Sciences),* 8, 51–86.

Kawano, S., Hiratsuka, A., & Hayashi, K. (1982). Life history characteristics and survivorship of *Erythronium japonicum. Oikos,* 38, 129–49.

Keeler, K. H. (1977). The extrafloral nectaries of *Ipomoea carnea* (Convolvulaceae). *American Journal of Botany,* 64, 1182–8.

– (1980). The extrafloral nectaries of *Ipomoea leptophylla* (Convolvulaceae). *American Journal of Botany,* 67, 216–22.

– (1981a). Function of *Mentzelia nuda* (Loasaceae) postfloral nectaries in seed defense. *American Journal of Botany,* 68, 295–9.

– (1981b). Infidelity by *Acacia*-ants. *Biotropica,* 13, 79–80.

Kennedy, J. S., & Fosbrooke, I. H. M. (1972). The plant in the life of an aphid. In *Insect–Plant Relationships,* ed. H. F. van Emden, pp. 129–40. Oxford: Blackwell Scientific Publications.

Kennedy, J. S., & Stroyan, H. L. G. (1959). Biology of Aphids. *Annual Review of Entomology,* 4, 139–60.

Kerner, A. (1878). *Flowers and Their Unbidden Guests.* London: Kegan Paul.

Kerner, A. von Marilaun, & Oliver, F. W. (1894). *The Natural History of Plants.* London: Blackie & Son.

Kevan, P. G., & Baker, H. G. (1983). Insects as flower visitors and pollinators. *Annual Review of Entomology,* 28, 407–54.

Kevan, P. G., Chaloner, W. G., & Savile, D. B. O. (1975). Interrelationships of early terrestrial arthropods and plants. *Palaeontology,* 18, 391–417.

Khalifman, I. A. (1961). *Use of ants for forest protection.* Lesnoe Khozyaistvo (Moscow), no. 2.

Kincaid, T. (1963). The ant-plant, *Orthocarpus pusillus* Bentham. *Transactions of the American Microscopical Society,* 82, 101–5.

King, T. J. (1977a). The plant ecology of ant-hills in calcareous grasslands. 1. Patterns of species in relation to ant-hills in southern England. *Journal of Ecology,* 65, 235–56.

– (1977b). The plant ecology of ant-hills in calcareous grasslands. 2. Succession on the mounds. *Journal of Ecology,* 65, 257–78.

– (1977c). The plant ecology of ant-hills in calcareous grasslands. 3. Factors affecting the population sizes of selected species. *Journal of Ecology,* 65, 279–315.

Kirchner, W. (1964). Jahreszyklische untersuchungen zur Reservestoffspeicherung und Uberlebensfahigkeit adulter Waldameisenarbeiterinnen. *Zoologische Jahrbucher (Physiologie),* 71, 1–72.

Kleinfeldt, S. E. (1978). Ant gardens: The interaction of *Codonanthe crassifolia* (Gesneriaceae) and *Crematogaster longispina* (Formicidae). *Ecology,* 59, 449–56.

Kloft, W., & Ehrhardt, P. (1959). Untersuchungen über Säugtätigkeit und Schadwirkung der Sitkafichtenlaus *Liosomaphis abietina* (Walk). *Phytopathology,* 35, 401–10.

Knox, R. B., Kenrick, J., Bernhardt, P., Marginson, R., Beresford, I., Baker, I., & Baker, H. G. (in press). Extrafloral nectaries as adaptations for bird pollination in *Acacia terminalis. American Journal of Botany.*

Knuth, P. (1906–9). *Handbook of Flower Pollination*. Oxford: Oxford University Press.

Koptur, S. (1979). Facultative mutualism between weedy vetches bearing extrafloral nectaries and weedy ants in California. *American Journal of Botany,* 66, 1016–20.

Koptur, S., Smith, A. R., & Baker, I. (1982). Nectaries in some neotropical species of *Polypodium* (Polypodiaceae): Preliminary observations and analyses. *Biotropica,* 14, 108–113.

Kraai, A. (1962). How long do honeybees carry germinable pollen? *Euphytica,* 11, 53–6.

Krombein, K. V. (1951). Wasp visitors of tuliptree honeydew at Dunn Loring, Virginia. *Annals of the Entomological Society of America,* 44, 141–3.

Kugler, C. (1980). The sting apparatus in the primitive ants *Nothomyrmecia* and *Myrmecia. Journal of the Australian Entomological Society,* 19, 263–7.

Kwan, S. C., Hamson, A. R., & Campbell, W. F. (1969). The effects of different chemicals on pollen germination and tube growth in *Allium cepa* L. *Journal of the American Society for Horticultural Science,* 94, 561–2.

Laine, K. J., & Niemela, P. (1980). The influence of ants on the survival of mountain birches during an *Oporinia autumnata* (Lep., Geometridae) outbreak. *Oecologia,* 47, 39–42.

Lande, R. (1976). Natural selection and random genetic drift in phenotypic evolution. *Evolution,* 30, 314–34.

Landis, B. J. (1967). Attendance of *Smynthurodes betae* (Homoptera: Aphididae) by *Solenopsis molesta* and *Tetramorium caespitum* (Hymenoptera: Formicidae). *Annals of the Entomological Society of America,* 60, 707.

Lange, R. (1967). Die Nährungsverteilung unter den Arbeiterinnen des Waldameisenstaates. *Zeitschrift für Tierpsychologie,* 24, 513–45.

Lanza, J., & Krauss, B. (1983). The ecological function of amino acids in extrafloral, artificial nectars. *Bulletin of the Ecological Society of America,* 64, 104.

Law, J. H., Wilson, E. O., & McCloskey, J. A. (1965). Biochemical polymorphism in ants. *Science,* 149, 543–6.

Leavitt, S. W., Dueser, R. D., & Goodell, H. G. (1979). Plant regulation of essential and non-essential heavy metals. *Journal of Applied Ecology,* 16, 203–12.

Leeper, G. W. (1949). *The Australian Environment C.S.I.R.O.,* Melbourne: University of Melbourne Press.

Leston, D. (1973). The ant mosaic-tropical tree crops and the limiting of pests and diseases. *Proceedings of the American National Academy of Sciences,* 19, 311–41.

– (1978). A neotropical ant mosaic. *Annals of the Entomological Society of America,* 71, 649–53.

Letourneau, D. K. (1983). Passive aggression: An alternative hypothesis for the *Piper–Pheidole* association. *Oecologia,* 60, 122–6.

Levieux, J. (1967). La place de *Camponotus acvapimensis* Mayr (Hymenoptera: Formicidae) dans la chaine alimentaire d'une savane de Cote-D'Ivoire. *Insectes Sociaux,* 14, 313–22.

Levieux, J., & Diomande, T. (1978). La nutrition des fourmis granivores. II. Cycle d'activité et régime alimentaire de *Brachyponera senaarensis* Mayr (Hymenoptera: Formicidae). *Insectes Sociaux,* 25, 187–96.

Levin, D. A., & Kerster, H. W. (1974). Gene flow in seed plants. *Evolutionary Biology,* 7, 139–220.

Levin, S. A. (1983). Some approaches to the modelling of coevolutionary interactions. In *Coevolution,* ed. M. H. Nitecki, pp. 21–65. Chicago: University of Chicago Press.

Levins, R., Pressick, M. L., & Heatwole, H. (1973). Coexistence patterns in insular ants. *American Scientist,* 61, 463–72.

Lieberman, D., Hall, J. B., Swaine, M. D., & Lieberman, M. (1979). Seed dispersal by baboons in the Shai Hills, Ghana. *Ecology,* 60, 65–75.

Likens, G. E., Bormann, F. H., Pierce, R. S., Eaton, J. C., & Johnson, N. M. (1977). *Biogeochemistry of a Forested Ecosystem.* New York: Springer.

Lin, N., & Michener, C. D. (1972). Evolution of sociality in insects. *Quarterly Review of Biology,* 47, 131–59.

Lingren, P. D., & Lukefahr, M. J. (1977). Effects of nectarless cotton on caged populations of *Campoletis sonorensis. Environmental Entomology,* 6, 586–8.

Linhart, L. B. (1973). Ecological and behavioral determinants of pollen dispersal in hummingbird-pollinated *Heliconia. American Naturalist,* 107, 511–23.

Llewellyn, M. (1972). The effects of the lime aphid, *Eucallipterus tiliae* L. (Aphidadae) on the growth of the lime, *Tilia* × *vulgaris* Hayne. *Journal of Applied Ecology,* 9, 261–82.

Llewellyn, M., Rashid, R., & Leckstein, P. (1974). The ecological energetics of the willow aphid *Tuberolachnus salignus* (Gmelin); honeydew production. *Journal of Animal Ecology,* 43, 19–29.

Lloyd, F. E. (1901). The extra-nuptial nectaries in the common brake, *Pteridium aquilinum. Science,* 23, 885–90.

Lofquist, J., & Bergstrom, G. (1980). Volatile communication substances in Dufour's gland of virgin females and old queens of the ant *Formica polyctena. Journal of Chemical Ecology,* 6, 309–20.

Longino, J. (1983). The influence of ants and butterflies on the growth of a neotropical liana, *Passiflora pittieri* (Mast). *Bulletin of the Ecological Society of America,* 64, 118.

Lu, K. L., & Mesler, M. R. (1981). Ant dispersal of a neotropical forest floor Gesneriad. *Biotropica,* 13, 159–60.

Ludwig, F. von (1899). Die Ameisen im Dienst der Pflanzenverbreitung. *Zeitschrift für Wissenschaftliche Zoologie,* 4, 38–41.

Luond, B., & Luond, R. (1981). Insect dispersal of pollen and fruits in *Ajuga. Candollea,* 36, 167–79.

Lyford, W. H. (1963). Importance of ants to brown podzolic soil genesis in New England. *Harvard Forest Paper,* no. 7, 1–17.

Madison, M. (1979). Additional observations on ant-gardens in Amazonas. *Selbyana,* 5, 107–15.

Majer, J. D. (1976a). The maintenance of the ant mosaic in Ghana cocoa farms. *Journal of Applied Ecology,* 13, 123–44.

– (1976b). The influence of ants and ant manipulation on the cocoa farm fauna. *Journal of Applied Ecology,* 13, 157–75.

– (1982). Ant-plant interactions in the Darling botanical district of Western Australia. In *Ant-Plant Interactions in Australia,* ed. R. C. Buckley, pp. 45–61. The Hague: Junk.

Malicky, H. von (1970). Unterschiede im Angriffsverhalten von *Formica*-arten (Hymenoptera, Formicidae) gegenüber Lycaenidenraupen (Lepidoptera). *Insectes Sociaux,* 17, 121–4.
Malozemova, L. A., & Koruma, N. P. (1973). Effect of ants on soil. *Ekologiya,* 4, 450–2.
Maltais, J. B., & Auclair, J. L. (1952). Occurrence of amino acids in the honeydew of the crescent-marked lily aphid, *Myzus circumflexus* (Buck.). *Canadian Journal of Zoology,* 30, 191–3.
Maramorosch, K. (1963). Arthropod transmission of plant viruses. *Annual Review of Entomology,* 8, 369–414.
Markin, G. P. (1970). Food distribution within laboratory colonies of the Argentine ant, *Iridomyrmex humilis* (Mayr). *Insectes Sociaux,* 17, 127–58.
Marshall, D. L., Beattie, A. J., & Bollenbacher, W. E. (1979). Evidence for diglycerides as attractants in an ant-seed interaction. *Journal of Chemical Ecology,* 5, 335–44.
Martin, A. C., Zim, H. S., & Nelson, A. L. (1951). *American Wildlife and Plants: A Guide to Wildlife Food Habits.* New York: Dover.
Martin, M. M., Boyd, N. D., Gieselmann, M. J., & Silver, R. G. (1975). Activity of faecal fluid of a leaf-cutting ant toward plant cell-wall polysaccharides. *Journal of Insect Physiology,* 21, 1887–92.
Maschwitz, U. (1974). Vergleichende Untersuchungen zur Funktion der Ameisenmetathorakaldrüse. *Oecologia,* 16, 303–10.
Maschwitz, U., Koob, K., & Schildknecht, H. (1970). Ein Beitrag zur Funktion der Metathorakaldrüse der Ameisen. *Journal of Insect Physiology,* 16, 387–404.
Maschwitz, U., Wüst, M., & Schurian, K. (1975). Bläulingsraupen als Zuckerlieferanten für Ameisen. *Oecologia,* 18, 17–21.
Masselink, A. K. (1980). Germination and seed population dynamics in *Melampyrum pratense* L. *Acta Botanica Neerlandica,* 29, 451–68.
Mattson, W. J. (1980). Herbivory in relation to plant nitrogen content. *Annual Review of Ecology and Systematics,* 11, 119–61.
Mattson, W. J., & Addy, N. D. (1975). Phytophagous insects as regulators of forest primary production. *Science,* 190, 515–22.
McClure, M. S. (1980). Foliar nitrogen: A basis for host suitability for elongate Hemlock scale, *Fiorinia externa* (Homoptera: Diaspididae). *Ecology,* 61, 72–9.
McCook, H. C. (1882). Ants as beneficial insecticides. *Proceedings of the Academy of Natural Sciences of Philadelphia,* pp. 263–71.
McDade, L. A., & Kinsman, S. (1980). The impact of floral parasitism in two neotropical hummingbird-pollinated plant species. *Evolution,* 34, 944–58.
McKey, D. (1975). The ecology of coevolved seed dispersal systems. In *Coevolution of Animals and Plants,* ed. L. E. Gilbert & P. H. Raven, pp. 159–91. Austin: University of Texas Press.
McNaughton, S. J. (1979). Grazing as an optimization process: Grass-ungulate relationships in the Serengeti. *American Naturalist,* 113, 691–703.
McNeil, J. N., Delisle, J., & Finnegan, R. J. (1977). Inventory of aphids on seven conifer species in association with the introduced red wood ant, *Formica lugubris* (Hymenoptera: Formicidae). *Canadian Entomologist,* 109, 1199–1202.

– (1978). Seasonal predatory activity of the introduced red wood ant *Formica lugubris* (Hymenoptera: Formicidae) at Valcartier, Quebec in 1976. *Canadian Entomologist,* 110, 85–90.

McNeil, S., & Southwood, T. R. E. (1978). The role of nitrogen in the development of insect/plant relationships. In *Biochemical Aspects of Plant and Animal Coevolution,* ed. J. Harborne, pp. 77–98. London: Academic Press.

Mehan, M., & Malik, C. P. (1975). Studies on the effect of different growth regulators on the elongation of pollen tubes in *Calatropis procera. Journal of Palynology,* 11, 71–7.

Meinwald, J., Smolanoff, J., Chibnall, A. C., & Eisner, T. (1975). Characterization and synthesis of waxes from homopterous insects. *Journal of Chemical Ecology,* 1, 269–74.

Melville, R. (1969). Leaf venation patterns and the origin of the angiosperms. *Nature,* 224, 121–5.

Messina, F. J. (1981). Plant protection as a consequence of an ant-membracid mutualism: Interactions on Goldenrod (*Solidago* sp.). *Ecology,* 62, 1433–40.

Meyer, C. R., & Meyer, V. G. (1961). Origin and inheritance of nectariless cotton. *Crop Science,* 1, 167–70.

Michener, C. D. (1974). *Social Behavior of Bees.* Cambridge, Mass.: Harvard University Press (Belknap).

Migliorato, E. (1910). Sull'impollinazione di *Rohdea japonica* Roth per mezzo delle formiche. *Annali di Botanica,* 8, 241–2.

Miles, P. W. (1968). Insect secretions in plants. *Annual Review of Phytopathology,* 6, 137–64.

Miles, P. W., & Lloyd, J. (1967). Synthesis of a plant hormone by the salivary apparatus of plant-sucking Hemiptera. *Nature,* 213, 801–2.

Milewski, A. V., & Bond, W. J. (1982). Convergence of myrmecochory in mediterranean Australia and South Africa. In *Ant–Plant Interactions in Australia,* ed. R. C. Buckley, pp. 89–98. The Hague: Junk.

Milton, K. (1979). Factors influencing leaf choice by howler monkeys: A test of some hypotheses of food selection by generalist herbivores. *American Naturalist,* 114, 362–78.

Mittler, T. E. (1958). Studies on the feeding and nutrition of *Tuberolachnus salignus* (Gmelin) (Homoptera: Aphididae). 2. The nitrogen and sugar composition of ingested phloem sap and excreted honeydew. *Journal of Experimental Biology,* 35, 74–84.

Monteith, L. G. (1967). Response by *Diprion hercyniae* (Hymenoptera: Diprionidae) to its food plant and their influence on its relationship with its parasite *Drino bohemica* (Diptera: Tachinidae). *Canadian Entomologist,* 99, 682–5.

Mothes, K. (1960). Alkaloids in the plant. In *The Alkaloids: Chemistry and Physiology,* ed. R. H. F. Manske, vol. 6, pp. 22–59. New York: Academic Press.

Motten, A. F., Campbell, D. R., Alexander, D. E., & Miller, H. L. (1981). Pollination effectiveness of specialist and generalist visitors to a North Carolina population of *Claytonia virginica. Ecology,* 62, 1278–87.

Muir, D. A. (1959). The ant-aphid-plant relationship in West Dunbartonshire. *Journal of Animal Ecology,* 28, 133–40.

Muller, H. (1883). *The Fertilization of Flowers.* London.

Muller, R. N. (1978). The phenology, growth and ecosystem dynamics of *Erythronium americanum* in the northern hardwood forest. *Ecological Monographs,* 48, 1–20.

Muller, R. N., & Bormann, F. H. (1976). The role of *Erythronium americanum* in energy flow and nutrient dynamics of a northern hardwood forest ecosystem. *Science,* 193, 1126–8.

Nakamura, S., Miki-Hirosige, H., & Iwanami, Y. (1982). Ultrastructural study of *Camellia japonica* pollen treated with myrmicacin, an ant-origin inhibitor. *American Journal of Botany,* 69, 538–45.

Nesom, G. L. (1981). Ant dispersal in *Wedelia hispida* HBK (Heliantheae: Compositae). *Southwestern Naturalist,* 26, 5–12.

Nesom, G. L., & Stuessy, T. F. (1982). Nesting of beetles and ants in *Clibadium microcephalum* S. F. Blake (Compositae: Heliantheae). *Rhodora,* 84, 117–24.

Nickerson, J. C., Kay, C. A. R., Buschman, L. L., & Whitcomb, W. H. (1977). The presence of *Spissistilus festinus* as a factor affecting egg predation by ants in soybeans. *Florida Entomologist,* 60, 193–9.

Nie, N. H., Hull, C. H., Jenkins, J. G., Steinbrenner, K., & Brent, D. H. (1975). *Statistical Package for the Social Sciences.* New York: McGraw-Hill.

Oates, J. F., Waterman, P. G., & Choo, G. M. (1980). Food selection by the south Indian leaf-monkey, *Presbytis johnii,* in relation to leaf chemistry. *Oecologia,* 45, 45–56.

O'Dowd, D. J. (1979). Foliar nectar production and ant activity on a neotropical tree, *Ochroma pyramidale. Oecologia,* 43, 233–48.

– (1980). Pearl bodies of a neotropical tree, *Ochroma pyramidale:* Ecological implications. *American Journal of Botany,* 67, 543–9.

– (1982). Pearl bodies as ant food: An ecological role for some leaf emergences of tropical plants. *Biotropica,* 14, 40–9.

O'Dowd, D. J., & Catchpole, E. A. (1983). Ants and extrafloral nectaries: No evidence for plant protection in *Helichrysum* spp.–ant interactions. *Oecologia,* 59, 191–200.

O'Dowd, D. J., & Hay, M. E. (1980). Mutualism between harvester ants and a desert ephemeral: Seed escape from rodents. *Ecology,* 61, 531–40.

Orians, G. H., & Janzen, D. H. (1974). Why are embryos so tasty? *American Naturalist,* 108, 581–92.

Orlob, G. B. (1963). The role of ants in the epidemiology of barley yellow dwarf virus. *Entomologia Experimentalis et Applicata,* 6, 95–106.

Osborne, D. (1972). Mutual regulation of growth and development in plants and insects. In *Insect-Plant Relationships,* ed. H. F. van Emden, pp. 33–42. Oxford: Blackwell Scientific Publications.

Oster, G. F., & Wilson, E. O. (1978). *Caste and Ecology in the Social Insects.* Princeton, N.J.: Princeton University Press.

Ovington, J. D. (1956). Studies of the development of woodland conditions under different trees. 5. The mineral composition of the ground flora. *Journal of Ecology,* 44, 597–604.

Percival, M. S. (1961). Types of nectar in angiosperms. *New Phytologist,* 60, 235–81.

– (1974). Floral ecology of coastal scrub in southeast Jamaica. *Biotropica,* 6, 104–29.

Peregrine, D. J., Mudd, A., & Cherrett, J. M. (1973). Anatomy and preliminary chemical analysis of the post-pharyngeal glands of the leaf-cutting ant, *Acromyrmex octospinosus* (Reich.) (Hymenoptera: Formicidae). *Insectes Sociaux,* 20, 355–63.

Petal, J. (1978). The role of ants in ecosystems. In *Production Ecology of Ants and Termites,* ed. M. V. Brian, pp. 293–325. Cambridge: Cambridge University Press.

Petal, J., Jakubczyk, H., & Wojcik, Z. (1967). L'influence des fourmis sur la modification des sols et des plantes dans le milieu des prairies. In *Methods of Study in Soil Ecology,* ed. J. Phillipson, pp. 235–40. Paris: Proceedings of the Paris Symposium, UNESCO.

Petersen, B. (1977). Pollination by ants in the alpine tundra of Colorado. *Transactions of the Illinois State Academy of Science,* 70, 349–55.

Pickett, C. H., & Clark, D. (1979). The function of extrafloral nectaries in *Opuntia acanthocarpa* (Cactaceae). *American Journal of Botany,* 66, 618–25.

Pierce, N. E., & Mead, P. S. (1981). Parasitoids as selective agents in the symbiosis between lycaenid butterfly larvae and ants. *Science,* 211, 1185–7.

Pimentel, D. (1955). Relationship of ants to fly control in Puerto Rico. *Journal of Economic Entomology,* 48, 28–30.

Pimentel, D., & Uhler, L. (1969). Ants and the control of houseflies in the Philippines. *Journal of Economic Entomology,* 62, 248.

Pisarski, B. (1978). Comparison of various biomes. In *Production Ecology of Ants and Termites,* ed. M. V. Brian, pp. 326–31. Cambridge: Cambridge University Press.

Plumstead, E. P. (1963). The influence of plants and environment on the developing animal life in Karroo times. *South African Journal of Science,* 59, 147–52.

Pontin, A. J. (1978). The numbers and distribution of subterranean aphids and their exploitation by the ant *Lasius flavus* (Fabr.). *Ecological Entomology,* 3, 203–7.

Porter, L. K., Viets, F. G., & Hutchinson, G. L. (1972). Air containing Nitrogen-15 Ammonia: Foliar absorption by corn seedlings. *Science,* 175, 759–61.

Price, P. W., Bouton, C. E., Gross, P., McPheron, B. A., Thompson, J. N., & Weis, A. E. (1980). Interactions among three trophic levels: Influence of plants on interactions between insect herbivores and natural enemies. *Annual Review of Ecology and Systematics,* 11, 41–65.

Proctor, M., & Yeo, P. (1973). *The Pollination of Flowers.* London: Collins.

Pudlo, R. J., Beattie, A. J., & Culver, D. C. (1980). Population consequences of changes in an ant-seed mutualism in *Sanguinaria canadensis. Oecologia,* 146, 32–7.

Putman, W. L. (1963). Nectar of peach leaf glands as insect food. *Canadian Entomologist,* 95, 108–9.

Quinlan, R. J., & Cherrett, J. M. (1979). The role of fungus in the diet of the leaf-cutting ant *Atta cephalotes* (L.). *Ecological Entomology,* 4, 151–60.

Raven, P. H. (1977). A suggestion concerning the Cretaceous rise to dominance of the angiosperms. *Evolution,* 31, 451–2.

Raven, P. H., Evert, R. F. & Curtis, H. (1981). *Biology of Plants,* 3d ed. New York: Worth.

Regal, P. J. (1977). Ecology and evolution of flowering plant dominance. *Science,* 196, 622–9.

Rehr, S. S., Feeny, P. P., & Janzen, D. H. (1971). Chemical defense in Central American non-ant-acacias. *Journal of Animal Ecology,* 42, 405–16.

Reyne, A. (1954). *Hippeococcus,* a new genus of Pseudococcidae from Java with peculiar habits. *Zoologische Mededelingen,* 32, 233–57.

Rhoades, D. F., & Cates, R. G. (1976). Toward a general theory of plant antiherbivore chemistry. In *Biochemical Interaction between Plants and Insects,* ed. J. W. Wallace & R. L. Mansell, pp. 168–213. New York: Plenum Press.

Rhyne, C. L. (1965). Inheritance of extrafloral nectaries in cotton. *Advances in the Frontiers of Plant Science,* 13, 121–35.

Richards, P. W. (1966). *The Tropical Rain Forest.* Cambridge: Cambridge University Press.

Rickson, F. R. (1969). Developmental aspects of the shoot apex, leaf and Beltian bodies of *Acacia cornigera. American Journal of Botany,* 56, 196–200.

– (1971). Glycogen plastids in Mullerian body cells of *Cecropia peltata* – a higher green plant. *Science,* 173, 344–7.

– (1973). Review of glycogen plastid differentiation in Mullerian body cells of *Cecropia peltata. Annals of the New York Academy of Sciences,* 210, 104–14.

– (1975). The ultrastructure of *Acacia cornigera* L. Beltian body tissue. *American Journal of Botany,* 62, 913–22.

– (1976). Anatomical development of the leaf trichilium and Mullerian bodies of *Cecropia peltata* L. *American Journal of Botany,* 63, 1266–71.

– (1977). Progressive loss of ant-related traits of *Cecropia peltata* on selected Caribbean islands. *American Journal of Botany,* 64, 585–92.

– (1979). Absorption of animal tissue breakdown products into a plant stem – the feeding of a plant by ants. *American Journal of Botany,* 66, 87–90.

– (1980). Developmental anatomy and ultrastructure of the ant food bodies (Beccariian bodies) of *Macaranga triloba* and *M. hypoleuca* (Euphorbiaceae). *American Journal of Botany,* 67, 285–92.

Rico-Gray, V. (1980). Ants and tropical flowers. *Biotropica,* 12, 223–4.

Ride, W. D. L. (1970). *A Guide to the Native Mammals of Australia.* Oxford: Oxford University Press.

Ridley, H. N. (1910). Symbiosis of ants and plants. *Annals of Botany,* 24, 457–83.

– (1930). *The Dispersal of Plants Throughout the World.* Ashford (UK): Reeve.

Risch, S. J., & Carroll, C. R. (1982). Effects of a keystone predaceous ant, *Solenopsis germinata,* on arthropods in a tropical agroecosystem. *Ecology,* 63, 1979–83.

Risch, S. J., & Rickson, F. R. (1981). Mutualism in which ants must be present before plants produce food bodies. *Nature,* 291, 149–50.

Risch, S. J., McClure, M., Vandermeer, J., & Waltz, S. (1977). Mutualism between three species of tropical *Piper* (Piperaceae) and their ant inhabitants. *American Midland Naturalist,* 98, 433–44.

Rockstein, M. (1978). *Biochemistry of Insects.* New York: Academic Press.
Rogers, L. L., & Applegate, R. D. (1983). Dispersal of fruit seeds by black bears. *Journal of Mammalogy,* 64, 310–11.
Rolfe, W. D. I., & Ingham, J. K. (1967). Limb structure, affinity and diet of the Carboniferous centipede *Arthropleura. Scottish Journal of Geology,* 3, 118–24.
Room, P. M. (1971). The relative distribution of ant species in Ghana's cocoa farms. *Journal of Animal Ecology,* 40, 735–51.
– (1973). Control by ants of pest situations in tropical tree crops: A strategy for research and development. *Papua New Guinea Agricultural Journal,* 24, 98–103.
Rosenthal, G. A., & Janzen, D. H. (1979). *Herbivores: Their Interactions with Secondary Plant Metabolites.* New York: Academic Press.
Ross, G. N. (1966). Life-history studies on Mexican butterflies. IV. The ecology and ethology of *Anatole rossi,* a myrmecophilous metalmark (Lepidoptera: Riodinidae). *Annals of the Entomological Society of America,* 59, 985–1004.
Russell, F. C. (1947). The chemical composition and digestibility of fodder shrubs and trees. *Joint Publications of the Commonwealth Agricultural Bureaux,* 10, 185–231.
Salisbury, E. J. (1942). *The Reproductive Capacity of Plants.* London: Bell.
Schaffer, W. M., Zeh, D. W., Buchmann, S. L., Kleinhans, S., Schaffer, M. V., & Antrim, J. (1983). Competition for nectar between introduced honeybees and native North American bees and ants. *Ecology,* 64, 564–77.
Schellner, R. A., Newell, S. J., & Solbrig, O. T. (1982). Studies on the population biology of the genus *Viola.* 4. Spatial pattern of ramets and seedlings in three stoloniferous species. *Journal of Ecology,* 70, 273–90.
Schemske, D. W. (1980). The evolutionary significance of extrafloral nectar production by *Costus woodsonii* (Zingiberaceae): An experimental analysis of ant protection. *Journal of Ecology,* 68, 959–67.
– (1982). Ecological correlates of a neotropical mutualism: Ant assemblages at *Costus* extrafloral nectaries. *Ecology,* 63, 932–41.
– (1983). Limits to specialization and coevolution in plant-animal mutualisms. In *Coevolution,* ed. M. H. Nitecki, pp. 67–109. Chicago: University of Chicago Press.
Scherba, G. (1962). Mound temperatures of the ant *Formica ulkei* Emery. *American Midland Naturalist,* 67, 373–85.
Schildknecht, H. (1976). Chemical ecology: A chapter of modern natural products chemistry. *Angewandte Chemie* (International Ed.), 15, 214–22.
Schildknecht, H., & Koob, K. (1970). Plant bioregulators in the metathoracic glands of Myrmicine ants. *Angewandte Chemie* (International Ed.), 9, 173.
– (1971). Myrmicacin, the first insect herbicide. *Angewandte Chemie* (International Ed.), 10, 124–5.
Schmitt, J. (1983). Density-dependent pollinator foraging, flowering phenology, and temporal pollen dispersal patterns in *Linanthus biedor. Evolution,* 37, 1247–57.
Schneider, P. (1972). Versuche zur Frage der individuellen Futterverteilung bei der kleinen Roten Waldameise (*Formica polyctena*). *Insectes Sociaux,* 19, 279–99.

Schneirla, T. C. (1971). *Army Ants: A Study in Social Organization.* San Francisco: Freeman.

Schubart, H. O. R., & Anderson, A. B. (1978). Why don't ants visit flowers? A reply to D. H. Janzen. *Biotropica,* 10, 310–11.

Scott, A. C. (1977). Coprolites containing plant material from the Carboniferous of Britain. *Paleontology,* 20, 59–68.

Scott, J. K. (1980). Ants protecting *Banksia* flowers from destructive insects? *Western Australian Naturalist,* 14, 151–4.

Sernander, R. (1906). Entwurf einer Monographie der Europäischen Myrmekochoren. *Kungl. Svenska Vetenskapsaka demiens Handlingar,* 41, 1–409.

Shea, S. R., McCormick, J., & Portlock, C. C. (1979). The effect of fires on regeneration of leguminous species in the northern jarrah (*Eucalyptus marginata* Sm.) forest of Western Australia. *Australian Journal of Ecology,* 4, 195–205.

Sheata, M. N., & Kaschef, A. H. (1971). Foraging activities of *Messor aegyptiacus* Emery (Hymenoptera: Formicidae). *Insectes Sociaux,* 18, 215–26.

Shukla, S. N., & Tewari, M. N. (1974). Antagonism between plant growth regulators in pollen tube elongation of *Calatropis procera. Experientia* (Basel), 30, 495.

Siccama, T. G., Bormann, F. H., & Likens, G. E. (1970). The Hubbard Brook ecosystem study: Productivity, nutrients, and phytosociology of the herbaceous layer. *Ecological Monographs,* 40, 389–402.

Simon, E. (1977). Cadmium tolerance in populations of *Agrostis tenuis* and *Festuca ovina. Nature,* 265, 328–30.

Skinner, G. J., & Whittaker, J. B. (1981). An experimental investigation of inter-relationships between the wood-ant (*Formica rufa*) and some tree-canopy herbivores. *Journal of Animal Ecology,* 50, 313–26.

Slansky, F., Jr., & Feeny, P. (1977). Stabilization of the rate of nitrogen accumulation by larvae of the cabbage butterfly on wild and cultivated food plants. *Ecological Monographs,* 47, 209–28.

Smallwood, J. (1982a). The effect of shade and competition on emigration rate in the ant *Aphaenogaster rudis. Ecology,* 63, 124–34.

– (1982b). Nest relocation in ants. *Insectes Sociaux,* 29, 138–47.

Smallwood, J., & Culver, D. C. (1979). Colony movements of some North American ants. *Journal of Animal Ecology,* 48, 373–82.

Smart, J., & Hughes, N. F. (1972). The insect and the plant: Progressive palaeoecological integration. In *Insect–Plant Relationships,* ed. H. F. van Emden, pp. 143–55. Oxford: Blackwell Scientific Publications.

Smirnov, B. A. (1962). Significance of ants in forest protection. *Zhashchita Rastenii ot Boleznei i Vreditelei,* no. 9.

Smirnov. V. I. (1966). Ants as a factor in forest protection against oak leaf roller. *Lesnoe Khozyaistvo* (Moscow), no. 2.

Smith, W. (1903). *Macaranga triloba:* A new myrmecophilous plant. *New Phytologist,* 2, 79–82.

Sogawa, K. (1982). The rice brown planthopper: Feeding physiology and host plant interactions. *Annual Review of Entomology,* 27, 49–73.

Sokal, R. R., & Rohlf, F. J. (1981). *Biometry,* 2d ed. San Francisco: Freeman.

Solbrig, O. T. (1981). Studies on the population biology of the genus *Viola*. 2. The effect of plant size on fitness in *Viola sororia*. *Evolution*, 35, 1080–93.

Solbrig, O. T., Jain, S., Johnson, G. B., & Raven, P. H. (1979). *Topics in Plant Population Biology*. New York: Columbia University Press.

Solbrig, O. T., Newell, S. J., & Kincaid, D. T. (1980). The population biology of the genus *Viola*. 1. The demography of *Viola sororia*. *Journal of Ecology*, 68, 521–46.

Soltanpour, P. N., Ludwick, A. E., & Reuss, J. (1979). *Guide to Fertilizer Recommendations in Colorado*. Fort Collins: Colorado State Extension Service.

Sorensen, A. A., Busch, T. M., & Vinson, S. B. (1983). Behaviour of worker subcastes in the fire ant, *Solenopsis invicta*, in response to proteinaceous food. *Physiological Entomology*, 8, 83–92.

Southwood, T. R. E. (1972). The insect-plant relationship: an evolutionary perspective. In *Insect-Plant Relationships*, ed. H. F. van Emden, pp. 3–30. Oxford: Blackwell Scientific Publications.

Soysa, S. W. (1940). Orchids and ants. *Orchidologia zeylanica*, 7, 88.

Sparks, S. D. (1941). Surface anatomy of ants. *Annals of the Entomological Society of America*, 34, 572–9.

Spradbery, J. P. (1973). *Wasps*. Seattle: University of Washington Press.

Spruce, R. (1908). *Notes of a Botanist on the Amazon and Andes*. Vol. II. London: Macmillan.

Stager, R. (1931). Ueber die Einwirkung van Duftstuffen und Pflanzenduften auf Ameisen. *Zeitschrift für wissenschaftliche Insektenbiologie*, 26, 55–65.

Stary, P. (1969). Aphid-ant-parasite relationship in Iraq. *Insectes Sociaux*, 16, 269–78.

Stebbins, G. L. (1974). *Flowering Plants: Evolution above the Species Level*. Cambridge, Mass.: Harvard University Press (Belknap Press).

Stebbins, G. L., & Hoogland, R. D. (1976). Species diversity, ecology and evolution in a primitive angiosperm genus: *Hibbertia* (Dilleniaceae). *Plant Systematics and Evolution*, 125, 139–54.

Steele, J. E. (1976). Hormonal control of metabolism in insects. *Advances in Insect Physiology*, 12, 240–307.

Stephenson, A. G. (1981). Iridoids deter nectar thieves from the flowers of *Catalpa speciosa*. *Bulletin of the Ecological Society of America*, 62, 165.

– (1982a). Iridoid glycosides in the nectar of *Catalpa speciosa* are unpalatable to nectar thieves. *Journal of Chemical Ecology*, 8, 1025–34.

– (1982b). The role of the extrafloral nectaries of *Catalpa speciosa* in limiting herbivory and increasing fruit production. *Ecology*, 63, 663–9.

Sterling, W. L. (1978). Fortuitous biological suppression of the boll weevil by the red imported fire ant. *Environmental Entomology*, 7, 564–8.

Stradling, D. J. (1978). Food and feeding habits of ants. In *Production Ecology of Ants and Termites*, ed. M. V. Brian, pp. 81–106. Cambridge: Cambridge University Press.

Strickland, A. H. (1947). Coccids attacking cacao (*Theobroma cacao* L.) in West Africa, with descriptions of five new species. *Bulletin Entomological Research*, 38, 497–523.

– (1951). The entomology of swollen shoot of cacao. 2. The bionomics and ecology of the species involved. *Bulletin Entomological Research,* 42, 65–103.
Strokov, V. V. (1956). *Techniques of using fauna for forest protections* (in Russian). Moscow: Goslesbumizdat.
Strong, F. E. (1963). Studies on lipids in some homopterous insects. *Hilgardia,* 34, 43–61.
– (1965). Detection of lipids in the honeydew of an aphid. *Nature,* 205, 1242.
Stryer, L. (1981). *Biochemistry,* 2d ed. San Francisco: Freeman.
Sudd, J. H. (1967). *An Introduction to the Behavior of Ants.* London: Arnold.
Svoboda, J. A., Thompson, M. J., Robbins, W. E., & Kaplanis, J. N. (1978). Insect steroid metabolism. *Lipids,* 13, 742–53.
Swain, T. (1977). The effect of plant secondary products on insect-plant coevolution. *Proceedings of the XV International Congress of Entomology,* pp. 249–56. College Park., Md.: Entomological Society of America.
– (1978). Plant-animal coevolution; A synoptic view of the Paleozoic and Mesozoic. In *Biochemical Aspects of Plant and Animal Coevolution,* ed. J. B. Harborne, pp. 3–18. London: Academic Press.
Swan, L. A. (1964). *Beneficial Insects.* Evanston, Ill.: Harper & Row.
Sweetman, H. L. (1958). *The Principles of Biological Control.* Dubuque, Iowa: Brown.
Takhtajan, A. (1969). *Flowering Plants: Origin and Dispersal.* Edinburgh: Oliver & Boyd.
Talbot, M. (1953). Ants of the old-field community on the Edwin S. George Reserve, Livingston County, Michigan. *Contributions from the Laboratory of Vertebrate Biology, University of Michigan,* 63, 1–13.
Tamm, C. O. (1956). Further observations on the survival and flowering of some perennial herbs. *Oikos,* 7, 273–92.
Taylor, B. (1977). The ant mosaic on cocoa and other tree crops in western Nigeria. *Ecological Entomology,* 2, 245–55.
Taylor, T. H. C. (1937). *The Biological Control of an Insect in Fiji.* London: Imperial Institute of Entomology.
Tempel, A. S. (1983). Bracken fern (*Pteridium aquilinum*) and nectar-feeding ants: A non-mutualistic interaction. *Ecology,* 64, 1411–22.
Tevis, L. (1958). Interrelations between the harvester ant *Veromessor pergandei* and some desert ephemerals. *Ecology,* 39, 695–704.
Thompson, J. N. (1981). Reversed animal-plant interactions: The evolution of insectivorous and ant-fed plants. *Biological Journal of the Linnean Society,* 16, 147–55.
– (1982). *Interaction and Coevolution.* New York: Wiley.
Tilman, D. (1978). Cherries, ants and tent caterpillars: Timing of nectar production in relation to susceptibility of caterpillars to ant predation. *Ecology,* 59, 686–92.
Trager, W. (1953). Nutrition. In *Insect Physiology,* ed. K. D. Roeder, pp. 350–86. New York: Wiley.
Trelease, W. (1879). Nectar, what it is, and some of its uses. In *Report upon Cotton Insects,* ed. J. H. Comstock, pp. 319–43. Washington, D.C.: USDA.
Treub, M. (1883). Sur le *Myrmecodia echinata* Gaudich. *Annales du Jardin Botanique de Buitenzorg,* 3, 129–59.

Turnbull, C. L. (1984). The dynamics of an association between *Viola nuttallii* Pursh. and its seed dispersers, *Myrmica discontinua* Weber and *Formica podzolica* Francoeur. Ph.D. thesis, Northwestern University.

Turnbull, C. L., Beattie, A. J., & Hanzawa, F. M. (1983). Seed dispersal by ants in the Rocky Mountains. *Southwestern Naturalist,* 28, 289–93.

Turnbull, C. L., & Culver, D. C. (1983). The timing of seed dispersal in *Viola nuttallii:* Attraction of dispersers and avoidance of predators. *Oecologia,* 59, 360–5.

Ulbrich, E. (1939). Deutsche Myrmekochoren. *Repertorium Specierum Novarum Regni Vegetabilis,* 67, 1–60.

Ule, E. (1902). Ameisengarten im Amazonasgebiet. *Botanisches Jahrbucher,* 30, 45–52.

– (1905). Blumengarten der Ameisen am Amazonenstrome. *Vegetationsbilder,* Ser. 3, pt. 1, 1–14.

– (1906). Ameisenflanzen. *Botanisches Jahrbucher,* 37, 335–52.

Uphof, J. C. T. (1942). Ecological relations of plants with ants and termites. *Botanical Review,* 8, 563–98.

van der Pijl, L. (1954). *Xylocopa* and flowers in the tropics. *Proceedings Koninklijke Nederlandse Akademie van Wetenschappen,* 57, 541–62.

– (1955). Some remarks on myrmecophytes. *Phytomorphology,* 5, 190–200.

– (1972). *Principles of Dispersal in Higher Plants.* New York: Springer.

Vanderplank, F. L. (1960). The bionomics and ecology of the red tree ant, *Oecophylla* sp. and its relationship to the coconut bug *Pseudotheraptus wayi* Brown (Coreidae). *Journal of Animal Ecology,* 29, 15–33.

van Emden, H. F. (1972). Aphids as phytochemists. In *Phytochemical Ecology,* ed. J. B. Harborne, pp. 25–43. London: Academic Press.

van Emden, H. F., & Way, M. J. (1972). Host plants in the population dynamics of insects. In *Insect–Plant Relationships,* ed. H. F. van Emden, pp. 187–200. Oxford: Blackwell Scientific Publications.

Van Vorhis Key, S. E., & Baker, T. C. (1982). Specificity of laboratory trail following by the Argentine ant, *Iridomyrmex humilis* (Mayr), to (z)-9-hexadecenal, analogs, and gaster extract. *Journal of Chemical Ecology,* 8, 1057–63.

Vinson, S. B. (1968). The distribution of an oil, carbohydrate and protein food source to members of the imported fire ant colony. *Journal of Economic Entomology,* 61, 712–14.

– (1976). Host selection by insect parasitoids. *Annual Review of Entomology,* 21, 109–23.

Vitousek, P. M., Gosz, J. R., Grier, C. C., Mellillo, J. M., Reiners, W. A., & Todd, R. L. (1979). Nitrate losses from disturbed ecosystems. *Science,* 204, 469–74.

Vogel, S. (1978). Nectarien und ihre ecologische Bedeutung. *Apidologie,* 8, 321–35.

von Wettstein, R. (1889). Ueber die Compositen der österreichisch-ungarischen Flora mit zuckerabschiedenden Hullschuppen. *Oesterreichische Akademie der Wissenschaften. Sitzungsberichte.* Part I. *Minerologie, Krystallographie, Botanik,* 97, 570–89.

Wagner, W. H., Jr. (1972). *Solanopteris brunei,* a little-known fern epiphyte with dimorphic stems. *American Fern Journal,* 62, 33–43.

Walker, R. B. (1954). The ecology of serpentine soils. II. Factors affecting plant growth on serpentine soils. *Ecology,* 35, 259–66.

Warburg, O. (1892). Ueber Ameisenpflanzen (Myrmecophyten). *Biologichnii Zbirnik L'vivs'kii Derzhavnii Universitet,* 12, 129–42.

Way, M. J. (1953). The relationship between certain ant species with particular reference to the biological control of the coreid *Theraptus* sp. *Bulletin of Entomological Research,* 44, 669–91.

– (1954). Studies on the association of the ant *Oecophylla longinoda* (Lat.) with the scale insect *Saissetia zanzibarensis* Williams (Coccoidae). *Bulletin of Entomological Research,* 45, 113–34.

– (1963). Mutualism between ants and honeydew-producing Homoptera. *Annual Review of Entomology,* 8, 307–44.

– (1968). Intra-specific mechanisms with special reference to aphid populations. In *Insect Abundance,* ed. T. R. E. Southwood. *Symposium of the Royal Entomological Society of London,* 4, 18–36.

Way, M. J., & Banks, C. J. (1967). Intra-specific mechanisms in relation to the natural regulation of numbers in *Aphis fabae* Scop. *Annals of Applied Biology,* 59, 189–205.

Weber, N. A. (1943). Parabiosis in neotropical "ant-gardens." *Ecology,* 24, 400–4.

– (1944). The neotropical coccid-tending ants of the genus *Acropyga* Roger. *Annals of the Entomological Society of America,* 37, 89–120.

– (1966). Fungus-growing ants. *Science,* 153, 587–604.

Wells, T. C. E., Sheail, J., Ball, D. F., & Ward, L. K. (1976). Ecological studies on the Porton ranges: Relationships between vegetation, soils and land-use history. *Journal of Ecology,* 64, 589–626.

Went, F. W., Wheeler, J., & Wheeler, G. C. (1972). Feeding and digestion in some ants (*Veromessor* and *Manica*). *Bioscience,* 22, 82–8.

Westoby, M., & Rice, B. (1981). A note on combining two methods of dispersal-for-distance. *Australian Journal of Ecology,* 6, 23–7.

Westoby, M., Rice, B., Shelley, J. M., Haig, D., & Kohen, J. L. (1982). Plants' use of ants for dispersal at West Head, New South Wales. In *Ant–Plant Interactions in Australia,* ed. R. C. Buckley, pp. 75–88. The Hague: Junk.

Wheeler, W. M. (1910). *Ants: Their Structure, Development and Behavior.* New York: Columbia University Press.

– (1914). The ants of the Baltic amber. *Schriften der Physikalisch-Oekonomischen Gesellschaft zu Königsberg,* 55, 1–142.

– (1921). A new case of parabiosis and the "ant gardens" of British Guiana. *Ecology,* 2, 89–103.

– (1942). Studies of neotropical ant-plants and their ants. *Bulletin of the Museum of Comparative Zoology,* Harvard, 90, 1–262.

Whittaker, R. H. (1954). The ecology of serpentine soils. IV. The vegetational response to serpentine soils. *Ecology,* 35, 275–88.

– (1975). *Communities and Ecosystems,* 2d ed. New York: Macmillan.

Wigglesworth, V. B. (1972). *The Principles of Insect Physiology,* 7th ed. London: Chapman & Hall.

- (1974) *Insect Physiology,* 7th ed. London: Methuen.
Wild, A. (1958). The phosphate content of Australian soils. *Australian Journal of Agricultural Research,* 9, 193–204.
Williams, E. G., Ramm-Anderson, S., Dumas, C., May, S. L., & Clarke, A. E. (1982). The effects of isolated components of *Prunus avium* L. styles on *in vitro* growth of pollen tubes. *Plants,* 156, 517–19.
Williams, G. C. (1975). *Sex and Evolution.* Princeton, N.J.: Princeton University Press.
Willis, J. C. (1966). *A Dictionary of the Flowering Plants and Ferns.* Cambridge: Cambridge University Press.
Willmer, P. B., & Corbet, S. A. (1981). Temporal and microclimatic partitioning of the floral resources of *Justicia aurea* amongst a concourse of pollen vectors and nectar robbers. *Oecologia,* 51, 67–78.
Willson, M. F., Anderson, P. K., & Thomas, P. A. (1983). Bracteal exudates in two *Cirsium* species as possible deterrents to insect consumers of seeds. *American Midland Naturalist,* 110, 212–14.
Wilson, E. O. (1963). The social biology of ants. *Annual Review of Entomology,* 8, 345–68.
- (1970). Chemical communication within animal species. In *Chemical Ecology,* ed. E. Sondheimer & J. B. Simeone, pp. 133–53. New York: Academic Press.
- (1971). *The Insect Societies.* Cambridge, Mass.: Harvard University Press (Belknap Press).
- (1975). *Sociobiology.* Cambridge, Mass.: Harvard University Press (Belknap Press).
Wilson, E. O., & Eisner, T. (1957). Quantitative studies of liquid food transmission in ants. *Insectes Sociaux,* 4, 157–66.
Wilson, E. O., Carpenter, F. M., & Brown, W. L. (1967). The first Mesozoic ants. *Science,* 157, 1038–40.
Wilson, E. O., Durlach, N. I., & Roth, L. M. (1958). Chemical releasers of necrophoric behavior in ants. *Psyche,* 65, 108–14.
Winkler, H. (1906). Beiträge zur Morphologie und Biologie tropischer Blüten und Fruchte. *Englers Botanische Jahrbuecher,* 38, 233–71.
Witherby, H. F., Jourdain, F. C. R., Ticehurst, N. F., & Tucker, B. W. (1952). *The Handbook of British Birds.* London: Witherby.
Woodwell, G. M., Whittaker, R. H., & Houghton, R. A. (1975). Nutrient concentrations in plants in the Brookhaven oak-pine forest. *Ecology,* 56, 318–32.
Wright, S. (1932). The roles of mutation, inbreeding, crossbreeding and selection in evolution. *Proceedings of the 6th International Congress of Genetics,* 1, 356–66.
- (1969). *Evolution and the Genetics of Populations.* Vol. 2, *The Theory of Gene Frequencies.* Chicago: University of Chicago Press.
Wyatt, R. (1981). Ant-pollination of the granite outcrop endemic *Diamorpha smallii* (Crassulaceae). *American Journal of Botany,* 68, 1212–17.
Wyatt, R., & Stoneburner, A. (1981). Patterns of ant-mediated pollen dispersal in *Diamorpha smallii* (Crassulaceae). *Systemaic Botany,* 6, 1–7.

Yokoyama, V. Y. (1978). Relation of seasonal changes in extrafloral nectar and foliar protein and arthropod populations in cotton. *Environmental Entomology,* 7, 799–802.

Young, A. M. (1980). Notes on foraging of the giant tropical ant *Paraponera clavata. Journal of the Kansas Entomological Society,* 53, 35–55.

– (1982). Giant neotropical ant *Paraponera clavata* visits *Heliconia pogonantha* flower bracts in premontane tropical rain forest. *Biotropica,* 10, 223.

Zoebelein, G. (1956a). Der Honigtau als Nahrung der Insekten. 1. *Zeitschrift für angewandte Entomologie,* 38, 369–416.

– (1956b). De Honigtau als Nahrung der Insekten. 2. *Zeitschrift für angewandte Entomologie,* 39, 129–67.

Index